Fuzzy Rule-Based Modeling with Applications to Geophysical, Biological and Engineering Systems

systems engineering series

series editor
A. Terry Bahill, University of Arizona

Engineering Modeling and Design
William L. Chapman, Hughes Aircraft Company
A. Terry Bahill, University of Arizona
A. Wayne Wymore, Systems Analysis and Design Systems

Fuzzy Rule-Based Modeling with Applications to Geophysical, Biological and Engineering Systems
András Bárdossy, University of Stuttgart
Lucien Duckstein, University of Arizona

Linear Systems Theory
Ferenc Szidarovszky, University of Arizona
A. Terry Bahill, University of Arizona

Model-Based Systems Engineering
A. Wayne Wymore, Systems Analysis and Design Systems

The Road Map to Repeatable Success: Using QFD to Implement Change
Barbara A. Bicknell, Bicknell Consulting, Inc.
Kris D. Bicknell, Bicknell Consulting, Inc.

System Engineering Planning and Enterprise Identity
Jeffrey O. Grady, JOG System Engineering

System Integration
Jeffrey O. Grady, JOG System Engineering

The Theory and Applications of Iteration Methods
Ioannis K. Argyros, Cameron University
Ferenc Szidarovszky, University of Arizona

Fuzzy Rule-Based Modeling with Applications to Geophysical, Biological and Engineering Systems

András Bárdossy
Lucien Duckstein

CRC Press
Boca Raton New York London Tokyo

Library of Congress Cataloging-in-Publication Data

Bárdossy, András.
  Fuzzy rule-based modeling with applications to geophysical,
biological and engineering systems / András Bárdossy and Lucien
Duckstein.
    p. cm.
  Includes bibliographical references and index.
  ISBN 0-8493-7833-8
  1. Mathematical models. 2. Fuzzy systems. I. Duckstein, Lucien.
  II. Title.
TA342.B37 1995
511'.8—dc20
                             95-8177
                               CIP

Preface

Traditionally, systems engineering has been pervaded by numerical models of physical phenomena, which, more often than not, are based on differential equations. While in some cases these equations do stem from physical laws, a significant trend in modeling has used the "black-box" concept. Given a set of input-output data, the problem is to find a best-fit function in a prescribed Parameterized class. This approach is interesting if the class of functions is general enough to match a large range of data sets, and if the best-fit algorithms are efficient enough. These two requirements are of course contradictory: linear models are simple but their field of applicability is far from covering the practical needs.

A significantly distinct view of systems modeling has emerged with the coming of age of Artificial Intelligence. Instead of using input-output numerical data as the point of departure of a modeling procedure, Artificial Intelligence suggests the use of expert knowledge as a more sensible approach. The symbolic models of that kind are made of a set of qualitative, verbally meaningful, conditional statements whose merit is to be semantically clear. It is the contrary of the back-box view. Unfortunately this type of model, if easily interpreted by non-specialists, has a lot of limitations. Being construed as human-originated, it lacks empirical underpinnings. Moreover, being symbolic, its validation is difficult to carry out since empirical data are generally numerical.

The fuzzy rule-based approach, based on Zadeh's fuzzy sets and first applied to control problems by Mamdani, and further advocated as a model-building tool by Sugeno, among others, is an attempt to reconcile the empirical rigor of traditional systems engineering techniques and the interpretability of Artificial Intelligence descriptions. A fuzzy rule can be viewed altogether as the flexible model of a verbal statement and as a numerical function with local validity. Fuzzy logic reasoning methods come down to generalized interpolation methods between local models. There are three reasons why the fuzzy rule-based approach is worth considering: first, the range of applicability of fuzzy rule-based models is large because a large class of non-linear functions can be described. In some cases, they are universal approximators. A second reason is that these models remain simple because each fuzzy model is composed of a set of simple local models along with procedures to go smoothly from one domain to another. A third reason is that they may remain verbally interpretable. This last point is the crucial one and is a most sought after feature of Artificial Intelligence models. It explains why fuzzy rule-based systems are potentially more promising than neural nets. Neural nets are universal approximators made of local models that are connected and remain opaque. On the contrary, fuzzy models may be a tool to go from a set of numerical input-output data over to a verbal description that describes the behavior of a system in qualitative terms. Conversely, the learning step of a fuzzy rule-based model can be significantly improved if a priori knowledge is available and supplied by experts, under the form of a collection of rules to be improved by empirical testing.

This book is an introduction to the fuzzy rule-based approach to systems modeling, written by two scientists who have significant experience in traditional engineering methods, both theoretical and practical. It reports on several years of work in various fields where numerical models are identified on the basis of numerical data, like time series analysis, soil water movement and even some aspects of medical diagnosis. The emphasis is not so much on the linguistic significance of fuzzy rules as on the capability of fuzzy rule-based models to account for empirically observed data. Beyond the merits of the proposed methods for modeling and exploiting fuzzy rules, the interest of the book lies in the comparative experiments between traditional models and fuzzy ones. The authors deserve congratulations for showing that in many cases, fuzzy rule-based models can challenge traditional methods in their own area of excellence, that is, model-fitting. This book is certainly a worthwhile and nonclassical contribution to the systems engineering literature and, because it is done by recognized specialists in engineering, it may speed up the lengthy recognition process of fuzzy set theory as a valuable tool for building numerical models of systems that remain linguistically interpretable.

Didier DUBOIS
Toulouse, January 1995

Acknowledgments

The authors are deeply indebted to the following individuals: Istvan Bogardi, with whom we started to develop fuzzy set applications in hydrology and water resources; Bijaya Shrestha, who provided technical, editorial and moral support; Ilona Pesti and Mitzi Austin who proofed the earlier versions of the manuscript over and over again; our spouses Andrea and Aloha and children, who saw us even less than usual during the preparation; Terry Bahill, series editor who gave efficient advice throughout, and students in systems and industrial engineering classes who pointed out errors in the manuscript drafts.

András Bárdossy
Lucien Duckstein

To Adrienn and Gergely

Contents

Introduction

The description of natural and manmade systems with the help of mathematical formulae has always been a goal in science. From the so-called "laws" of physics of the 18th century which are really models of observed phenomena, to complex central or ecosystem models of the 1990's, one is lead to find a balance between mathematical accuracy and realism of models (Casti, 1993; Wymore, 1993). The ideal model would be simple and transparent, and it would satisfy the principle of parsimony of parameters and at the same time be representative of reality (Bernier, 1994). Models operate in an idealized world, but unfortunately reality may be intricate and fraught with imprecision and uncertainties. We propose to recognize explicitly these characteristics of real systems by using fuzzy logic.

Uncertainty has long been considered a removable artifact which should gradually disappear with increasing knowledge. However, in the last decades it has been recognized that uncertainty, imprecision, and ambiguity are inevitable and probably inherent parts of natural systems. The traditional way of dealing with uncertainty — the application of probability theory and statistics — was successful in a great number of cases. However, the use of these theories for complex models requires a great number of theoretical hypotheses and often becomes extremely difficult and unusable for practical purposes. Overemphasized precision however increases the complexity of models, sometimes to a degree where their handling becomes impossible. Further precision does not always mean greater truth. Fuzzy set logic provides another means for dealing with uncertainty. As it deals with imprecise objects, it has been and, for a number of scientists, remains an unacceptable tool in the precise world of science. The success of fuzzy logic thus surprisingly began with industrial applications including train control (Yasunobu and Miyamoto 1985), auto focusing cameras (Shingu and Nishimori 1989), or cement kiln control (Holmblad and Ostergaard 1982).

The purpose of this monograph is thus to provide an interdisciplinary framework to model, by means of fuzzy rules, complex systems having imprecise, vaguely defined, or uncertain elements. The methodology developed herein can also be applied to an approximate model of well defined and exact systems. It can be applied both for the assessment of models and for the simplification of complex models. We are striving to show that fuzzy rule-based models are usable in both "soft" disciplines, such as ecology or medicine, where no generally accepted and precise "laws" are available for modeling purposes, and "hard" disciplines, such as physics and engineering, with accepted mathematical models. The fuzzy rule-based models can easily be coupled — for example, a "hard" model for flow in porous media may be coupled with a "soft" bacteriological growth model. Furthermore, the level of detail of models may be varied (by changing the rule system) without having to redesign the model.

Rule-based modeling has formed the essence of so-called expert systems (Zimmermann, 1985; Bahill, 1991); however, fuzzy rules have not quite been used in the same manner as herein. Fuzzy rules form the basis of fuzzy control systems, which have been and are used in many practical applications, especially industrial ones in Japan and Europe (Sugeno and Yasukawa, 1993). Industrial success stories of fuzzy control include portable video cameras, automatic transmission of automobiles, furnace temperature, robotics, urban underground railway, and banking. Yet all these fuzzy control systems include feedback which generally can provide guidance for adjusting the rules, such a procedure is in contrast with the open-loop modeling similar to system identification that we are developing in this research monograph.

Why fuzzy rules? Human way of thinking often appears to be based on rules of the type:

If events A and B or A and C and D occur, then the consequences may be either E or F and G.

rather than on the concept of a function which associates with every point in the domain a unique point in the range. However the above formulation is too vague to perform computations with the model. Fuzzy rule systems provide a means to translate the above natural statements into a computationally usable form. In principle it is possible to express a rule system directly by means of multidimensional functions defined on a multidimensional space, but then we lose the advantages of simplicity and visualization.

Pure or hard sciences strive to be exact, expressing the behavior of physical systems by means of "laws" encoded for instance as partial differential equations or complicated dynamical systems; an example of the latter is meteorological modeling for weather forecasting. Reality is even more complex than the models considered in science (Casti, 1990, 1993). Data availability is often insufficient to calibrate and validate complex models, and then heterogeneity or non-stationarity can distort results obtained by analytical or numerical modeling (Duckstein and Parent, 1994).

On the other hand, disciplines such as social and biological sciences, ecology, and medicine express their knowledge mostly in the form of connections and rules rather than in an explicit mathematical form (Casti et al., 1979). Interdisciplinary modeling may have thus been hampered by the major stumbling block of linking mathematical models built in the exact or "hard" sciences with the connections and verbal rule-based modeling used in the less exact or "soft" sciences. For example, biological modeling requires chemistry, an exact science; conversely, water resources decision-making leans on the hard science of hydrology but requires knowledge of social and institutional factors so as to be able to evaluate the consequences of decisions (Waterstone, 1994).

Our experience to date shows that the fuzzy rule-based approach has wide applicability, as illustrated by the real life examples in geophysics (especially hydrology), medical sciences, ecology, and economics, which are presented in this monograph. The method is easy to use, transparent, and robust. Clearly it is less accurate than pure mathematical modeling which, anyway, can only be applied (usually) to idealized or even "sterile" problems. In the case of heterogeneous real-life problems, a purely mathematical or analytical modeling can often only give a false impression of accuracy. Numerical accuracy does not mean that the model really describes the phenomenon considered. The gain in simplicity, computational speed, and flexibility that results from adopting to a fuzzy rule-based modeling procedure may compensate for the possible loss in accuracy. Furthermore, the size of a fuzzy rule set (that is, the number of rules used) may be adjusted to match the amount and accuracy of input data.

Fuzzy rule-based methods are often compared and sometimes even combined with neural nets. There are a lot of similarities but also differences between these methods. They are both able to describe the nonlinear behavior of systems. However, in contrast to the black box type NN, the fuzzy rules are very transparent as rules are explicitly stated. On the other hand, NN-s offer good training possibilities, which sometimes give the impression that the user has nothing to do but to put his data in and a model is produced. This is not true, and in NN-s one often faces the problem of "over learning" which means the network learns special details instead of the essential. This problem can be avoided using fuzzy rule-based systems. There are more and more algorithms available to assess rule systems (including control systems) from data, also simplifying their use. In any case, theoretical results on combination methods and rule systems presented in this book can also be useful for practitioners in fuzzy control.

This book is only meant to describe fuzzy rule-based systems. The interested reader should consult Dubois and Prade (1980a) or Zimmermann (1985) for a comprehensive treatment of fuzzy set theory.

The book is organized as follows:

Chapter 2 contains all basic definitions of fuzzy set theory used in this book. It should enable the reader without any prior knowledge on fuzzy sets to follow the subsequent chapters.

Chapter 3 provides a short introduction to fuzzy logic. The definition of fuzzy rules is given and different rule combination methods are introduced; several defuzzification methods are then developed.

Chapter 4 describes and defines fuzzy rule systems, and investigates rule response functions. It also gives information on the problems of selection of membership functions, inference and defuzzification methods.

Chapter 5 describes methods for the assessment of rule systems using expert knowledge and data. Verification and redundancy issues are also treated.

Chapter 6 briefly defines and illustrates fuzzy rule-based control. It compares fuzzy rule-based control to fuzzy rule-based modeling.

Chapter 7 describes fuzzy rule systems with discrete response sets. An atmospheric circulation pattern classification example illustrates the methodology.

Chapter 8 describes the use of fuzzy rule systems for modeling time series. A water demand forecasting example illustrates this type of problem.

Chapter 9 describes the use of fuzzy rule systems for exact physical systems described by partial differential equations. An example of soil water movement in the unsaturated zone illustrates the methodology.

Chapter 10 describes further applications, including a medical decision model concerning hypertension, and a sustainable reservoir operation example.

Appendix A contains the precise statements of selected propositions stated in the text and their proofs for the mathematically minded readers.

Basic elements and definitions

The goal in modeling systems is to find and to formulate governing laws in the form of precise mathematical terms. However, it has long been recognized that such perfect descriptions are not always possible. Incomplete and imprecise knowledge, observations that are often of a qualitative nature, and the great heterogeneity of the surrounding world cause many uncertainties in modeling. There are several possibilities to take such uncertainties into account. The evolution of frequentist probability theory and classical statistics has provided theoretical means for the description of uncertainty. However, these theories are based on frequencies repeated trials. This is not always the case in our unique world. The application of stochastic methods modeling usually requires much effort. Furthermore, a great number of assumptions have to be made. The application of these methods — such as regression techniques — is widely accepted, although, often the underlying assumptions aren't even stated (or sometimes unknown to the modeler).

An alternative to frequentist probability and statistics is a Bayesian approach, in which subjective knowledge may be substituted for repeated trials (Berger, 1985). Still, the application of Bayesian techniques necessitates many assumptions, often burdensome computations, and (most of all) represents a different philosophical approach to uncertainty analysis (Klir and Folger, 1988; Duckstein, 1994).

Fuzzy sets are used to describe uncertainty and imprecision in a non-probabilistic (non-frequentist) framework. Fuzzy sets were first introduced in Zadeh (1965), and have been applied in various fields, such as decision making and control. The purpose of this chapter is to provide a brief review of the definitions of fuzzy sets, linguistic variables, fuzzy numbers and fuzzy operations. There are quite a few books containing further details on fuzzy sets and fuzzy arithmetic such as Dubois and Prade (1980a), Zimmermann (1985), or Kaufmann and Gupta (1991).

2.1 Fuzzy sets: definitions and properties

2.1.1 Membership functions

A fuzzy set is a set of objects without clear boundaries or without well-defined characteristics. In contrast with ordinary sets where for each object it can be decided whether it belongs to the set or not, a partial membership in a fuzzy set is possible. An example of a fuzzy set could be "the set of long streets in Berlin". There are streets which clearly belong to the above set, and others which cannot be considered as long. But, if the concept of long is not exactly defined (for example ≥ 1700 m), there is a certain "gray" zone where the judgment outcome is not obvious (somewhat long streets). An exact definition of a long street would not conform with the human way of thinking. Why should a street of a length of 1701 m be long and an other one of 1699 m not be so? The normal human observer cannot perceive the difference — only measurements could decide on the membership. However, in real life one can well use the concept of long and short without taking any measurements.

Formally a fuzzy set (a subset really) is defined as follows:

Definition 2.1 *Let X be a set (universe). A is called a fuzzy subset of X if A is a set of ordered pairs:*

$$A = \{(x, \mu_A(x)); x \in X \, \mu_A(x) \in [0,1]\}$$

where $\mu_A(x)$ is the grade of membership of x in A. The function $\mu_A(x)$ is called the membership function of A.

The closer $\mu_A(x)$ is to 1 the more x is considered to belong to A — the closer it is to 0 the less it is taken as belonging to A. If $[0,1]$ is replaced by the two-element set $\{0,1\}$, then A can be regarded as an ordinary subset of X.

There are two basic forms in which a fuzzy set can be described:

1. By utilizing the pairwise form as given in the definition. This description form is used for fuzzy sets defined on a discrete set X. In this case, the elements of X are listed together with their membership values. For example, if $X = \{1, 2, 3\}$ than $A = \{(1, 0.1); (2, 1); (3, 0.7)\}$ is a fuzzy subset of X, with 1 having the membership value 0.1, 2 having the membership value 1, and 3 having the membership value 0.7.

2. By defining the membership function $\mu_A(x)$ directly. Usually this is the method applied if X is not a discrete set.

In this text, for simplicity, we use the phrase "fuzzy set" instead of "fuzzy subset". The set X is only specified explicitly if it is not obvious.

The membership function $\mu(x)$ can be regarded as the generalization of the characteristic function $\chi(x)$ of ordinary subsets of X defined as

$$\chi_A(x) = \begin{cases} 1 & \text{if } x \in A \\ 0 & \text{else} \end{cases}$$

Using the characteristic function as membership function each proper subset of X can also be regarded as a fuzzy subset.

Example 2.1 *The set of young persons is fuzzy, as there is no generally accepted boundary between young and not young. One might say that a person younger than 25 is certainly young, and older than 40 is certainly not young. Thus the membership function of this set A may be defined as*

$$\mu_A(x) = \begin{cases} 1 & \text{if } x \leq 25 \text{ years} \\ \frac{40-x}{15} & \text{if } 25 < x \leq 40 \\ 0 & \text{if } 40 < x \end{cases}$$

Figure 2.1 shows the graph of the corresponding membership function. Note that the definition of young persons depends on the environment

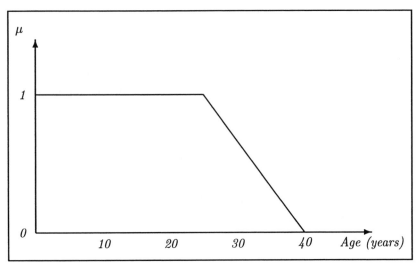

Figure 2.1. Membership function of a young person

in which it is used. For example in an elementary school, young and old would have a totally different fuzzy definition from the one just given.

In the transition zone between surely young, and surely not young a linear membership function was assumed. Another function could have also be selected, such as it will be done for the case of L-R fuzzy numbers introduced later in this chapter.

The membership value $\mu_A(x)$ can also be regarded as the "truth value" of the statement x belongs to A or, equivalently, the degree of fulfillment of the statement defining A, such as "young person". In some fields, especially scientific ones, there is a tendency to define sets with sharp boundaries and to accept only "true" or "not true" statements. However, in our everyday actions, we use imprecise terms and partial truth. In example 2.1 the statement "a 31-year-old person is young" has a truth value of $\mu_A(31) = 0.6$. As another example, we feel that the statement of a child calling a telephone booth a house is more true than if he or she is calling a dog a house. In the classical binary logic both statements would be false without any nuance. Learning may be considered a process that gradually removes fuzziness and replaces it by more and more precise definitions.

To compare and combine fuzzy sets and, later, fuzzy rules, it is useful to generalize the notion of set cardinality to discrete fuzzy sets.

Definition 2.2 *The cardinality of the fuzzy set* $A = \{(a_1, \mu_1), \ldots, (a_I, \mu_I)\}$ *is*

$$car(A) = \sum_{i=1}^{I} \mu_i$$

The cardinality of A is thus a real number between 0 and I.

Elements of a fuzzy set with a membership of at least a given value h form the (ordinary) set called h-level set:

Definition 2.3 *The h level set of a fuzzy set A is the set*

$$A(h) = \{x \; ; \; \mu_A(x) \geq h\}$$

Verbally this means the h-level set of the fuzzy set A consists of all elements that fulfill the definition of A to at least a degree h. By the definition:

$$A(h_1) \subset A(h_2) \quad \text{if} \quad h_1 > h_2$$

The h-level sets of A $(0 < h \leq 1)$ fully define the fuzzy set A. In practice, it is sufficient to calculate a few h-level sets to obtain an approximation of fuzzy set A.

2.1.2 Operations on fuzzy sets

Basic set theoretical operations can also be defined on fuzzy sets. There are several possibilities to define the intersection and the union of fuzzy sets. The classical ones suggested by Zadeh (1965) are as follows:

Definition 2.4 *The membership function of the complement* $C = A^c$ *of the fuzzy set* A *is defined by:*

$$\mu_C(x) = 1 - \mu_A(x)$$

Definition 2.5 *The membership function of the intersection* $D = A \cap B$ *of two fuzzy sets* A *and* B *is defined by:*

$$\mu_D(x) = \min\left(\mu_A(x), \mu_B(x)\right)$$

Definition 2.6 *The membership function of the union* $E = A \cup B$ *of two fuzzy sets* A *and* B *is defined by:*

$$\mu_E(x) = \max\left(\mu_A(x), \mu_B(x)\right)$$

These operations are simple generalizations of the usual set theoretic operations. For ordinary subsets of X with the characteristic function taken as membership function the above defined operations yield the usual intersection and union. For this definition the intersection/union of the h-level sets of the fuzzy sets equals the h-level set of the intersection/union. Figure 2.2 shows the union and Figure 2.3, shows the intersection of two fuzzy sets.

Note that the intersection of a fuzzy set A with its complement A^c is not necessarily empty. The union of these two sets is not always the universal set X as it would be in the case of ordinary sets. This property has often been criticized — however, it does not contradict our intuition: Consider a person having light brown hair. He or she cannot be called blond — but the statement that this person is not blond is also not true. So, for this person, being blond or not being blond cannot be decided with certainty.

Besides the minimum and maximum operations as given above, there are many functions that can be used to define operations, including intersection and union on fuzzy sets. A fundamental reason for using other functions is that the maximum and minimum values only depend on one of the two arguments so that the operation yields the same value in a

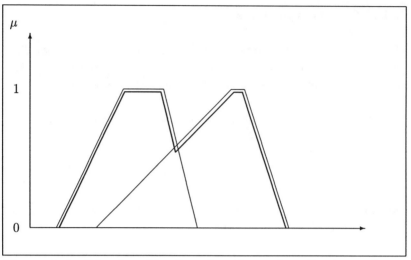

Figure 2.2. Membership function the union C of two fuzzy sets A and B (thick line)

range of the other argument that may be quite large in practice. However, we often feel that the statement:

$$(x \text{ is } A) \text{ and } (x \text{ is } B)$$

Figure 2.3. Membership function the intersection D of two fuzzy sets A and B (thick line)

is more "true" than $(x$ is $A)$ and $(x$ is $C)$ if $(x$ is $B)$ is more "true" than $(x$ is $C)$. To illustrate this suppose that $\mu_A(x) = \mu_B(x) = 0.1$ and $\mu_C(x) = 0.9$. In this case the intersection as defined above yields:

$$\mu_{A \cap B}(x) = \min(0.1, 0.1) = \mu_{A \cap C}(x) = \min(0.1, 0.9) = 0.1$$

which contradicts our intuitions. To overcome this problem functions for the definition of the intersection and the union based on the concepts of t-norms and t-conorms can be used. We thus provide definitions of t-norms and t-conorms, so as to construct a proper framework for the union and intersection operations on fuzzy sets.

The intersection of fuzzy sets can be defined with a so called t-norm function. This function assigns the membership value for an element in the intersection depending on the individual membership values in the intersecting sets.

Definition 2.7 *A t-norm is a bivariate function:*

$$t \; : \; [0,1] \times [0,1] \rightarrow [0,1]$$

with the following properties:

1. $t(0,0) = 0$
2. $t(x,1) = x$
3. $t(u,v) \leq t(w,z)$ *if* $u \leq w$ *and* $v \leq z$ *(Monotonicity)*
4. $t(x,y) = t(y,x)$ *(Symmetry)*
5. $t(x,t(y,z)) = t(t(x,y),z)$ *(Associativity)*

A t-norm can be taken as an intersection operator - applied to the membership functions of A and B. Let $D = A \cap B$, then:

$$\mu_D(x) = t\left(\mu_A(x), \mu_B(x)\right)$$

The properties of a t-norm describe the natural properties of an intersection:

The first property states that if an element does not belong to either one of the sets, than it does not belong to the intersection.

The second property means that an element that surely belongs to one of the sets belongs to the intersection at the same level as it belongs to the other set.

The third property is the above mentioned monotonicity: if an element belongs to both intersecting sets more than another one, then it also belongs to the intersection more than the other one.

Properties four and five assure that the intersection is independent of the order in which the intersecting sets are considered.

It can be seen that an intersection defined by applying a t-norm to ordinary subsets of X defined by their characteristic functions yields the ordinary intersection of these sets. Thus t-norms define generalizations of the intersection of ordinary sets.

The union of fuzzy sets can be defined with a so called t-conorm function. This function assigns the membership value to an element in the union depending on the individual membership values in their respective sets.

Definition 2.8 *A t-conorm is a bivariate function:*

$$c \; : \; [0,1] \times [0,1] \to [0,1]$$

with the following properties:

1. $c(1,1) = 1$
2. $c(x,0) = x$
3. $c(u,v) \le c(w,z)$ *if* $u \le w$ *and* $v \le z$ *(Monotonicity)*
4. $c(x,y) = c(y,x)$ *(Symmetry)*
5. $c(x,t(y,z)) = c(t(x,y),z)$ *(Associativity)*

A t-conorm can be taken as a union operator - applied to the membership functions of A and B. Let $E = A \cup B$, then:

$$\mu_E(x) = c\left(\mu_A(x), \mu_B(x)\right)$$

As with the t-norm taken as the intersection operator, the properties of the t-conorms ensure that the union is independent of the order in which the arguments are taken. The union so defined is also a generalization of the union of ordinary sets.

There is a close relationship between t-norms and t-conorms as shown in Schweizer and Sklar (1961); namely if t is a t-norm then

$$c(x,y) = 1 - t(1 - x, 1 - y) \tag{2.1.1}$$

defines a t-conorm.

Vice versa, if c is a t-conorm then:

$$t(x,y) = 1 - c(1 - x, 1 - y) \tag{2.1.2}$$

defines a t-norm.

There are several forms of t-norms suggested in the literature. Table 2.1 lists a few of these with their corresponding t-conorms. These norms,

Table 2.1. **t-norms and corresponding t-conorms**

	t-norm $t(x,y)$	t-conorm $c(x,y)$
Algebraic product-sum	xy	$x + y - xy$
Hamacher product-sum	$\frac{xy}{x+y-xy}$	$\frac{x+y-2xy}{1-xy}$
Einstein product-sum	$\frac{xy}{1+(1-x)(1-y)}$	$\frac{x+y}{1+xy}$
Bounded difference-sum	$\max(0, x + y - 1)$	$\min(1, x + y)$
Dubois and Prade (1980b) operators $(0 \leq p \leq 1)$	$\frac{xy}{\max(x,y,p)}$	$1 - \frac{(1-x)(1-y)}{\max[(1-x),(1-y),p]}$

which can all be used for defining the intersection and union of fuzzy sets, are seen to be a function of both independent variables, in contrast with the maximum and minimum operators. The choice of a norm depends both on the application and the preference of the modeler as explained, for example, in Dubois and Prade (1980b). In our experience, we have found that the algebraic product-sum provides a good alternative and is computationally parsimonious.

2.1.3 Linguistic variables

Besides numbers and objects defined in natural language, words, phrases, or expressions are also used to describe different phenomena. Much of our experience and knowledge is in a linguistic form which practically is not usable for computers requiring precise (usually numeric) information. The task of translating the imprecise verbal information into a computer usable numerical form can be performed using fuzzy sets. Linguistic variables and fuzzy sets are not the same — fuzzy sets are used to express the contents of a linguistic variable. A linguistic variable is a variable with values that are words or sentences. Formally a linguistic variable has a rather complicated definition (Zadeh, 1975):

Definition 2.9 *A linguistic variable, y is a 4-tuple (T, X, G, M) where T is a set of natural language terms from which x can take on its values, X is a universe, a common set on which the fuzzy sets corresponding to the linguistic variable are defined, G is a (context free) grammar used to generate the elements of T, and M is the mapping from T to the fuzzy subsets of X.*

This formal definition can be illustrated with the following example:

Example 2.2 *The height of an adult person is given as numerical value on the universe $X = [0, 300]$ (cm). The possible values for the linguistic variable are:*
$T = $ tall + very tall + not tall + small + very small + normal + rather small ...
For each of these expressions a corresponding fuzzy set on X is given by the mapping M. For example, M assigns the fuzzy set with the membership function:

$$\mu_A(x) = \begin{cases} 0 & \text{if } x \leq 170 \text{ cm} \\ \frac{185-x}{15} & \text{if } 170 < x \leq 185 \\ 1 & \text{if } 185 < x \end{cases}$$

to the element "tall" from T.

Sometimes there is no set X which can be naturally associated to the linguistic concept. Consider, for example, the linguistic terms such as good, excellent environmental quality. A set on which these concepts are measured also has to be defined.

2.1.4 Linguistic modifiers

In natural languages a specification of the properties is often done using linguistic modifiers. These modifiers (hedges) might both decrease and increase the uncertainty. Some of these hedges are:

VERY	FAIRLY	MOSTLY
OFTEN	SOMEWHAT	INDEED
ROUGHLY	ALMOST	MORE OR LESS
SORT OF	PRACTICALLY	NOT
MOST OF	AT LEAST A FEW	

These hedges are applied to vague concepts, resulting in either a more precise or imprecise (vague) description. It is interesting to observe that these hedges are often applied to very "precise" terms — softening them and making them acceptable or just plain understandable. Even results of scientific investigations are often communicated using linguistic hedges for a better understanding. These hedges can be replaced by operators acting on the membership function of the corresponding "unhedged" expression. For example a typical modificator is the hedge VERY:

$$\mu_{\text{VERY}}(x) = \mu(x)^2 \qquad (2.1.3)$$

As $x^2 < x$ for $0 < x < 1$ the above operator decreases the membership of the "uncertain" elements of the fuzzy set. This operator is often called concentration.

Another operator is the dilation, which is the opposite of the concentration and corresponds to the hedge MORE OR LESS, and may be taken as:

$$\mu_{\text{MORE OR LESS}}(x) = \sqrt{\mu(x)} \qquad (2.1.4)$$

This operator increases the membership of uncertain elements.

The hedge INDEED may be defined as:

$$\mu_{\text{INDEED}}(x) = \begin{cases} 2\mu(x)^2 & \text{if } 0 \leq \mu(x) < 0.5 \\ 1 - 2(1 - \mu(x))^2 & \text{if } 0.5 \leq \mu(x) \leq 1 \end{cases} \qquad (2.1.5)$$

This operator increases the contrast — elements having membership functions above 0.5 are assigned greater truth value (membership function), and elements whose membership function is below 0.5 are assigned smaller membership values. Figure 2.4 shows the application of hedges to the membership function corresponding to tall.

Operators for other linguistic hedges can be defined in a similar way. As a result, natural language statements can easily be converted into fuzzy sets.

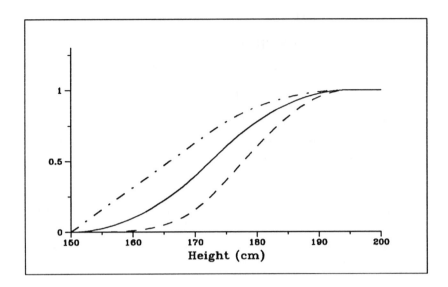

Figure 2.4. Membership function of **TALL** (solid), **VERY, TALL** (dashed), and **MORE OR LESS TALL** (dashed – dotted).

Linguistic variables make a natural language computation possible. After such computations have taken place it is desired to translate the results back to natural language statements. For this purpose the resulting fuzzy set membership function is compared to membership functions of given statements and the closest one is selected. An extension of the usual Euclidean distance is often used as a measure of closeness. Schmucker (1984) presents an example of natural language computation for risk analysis.

2.2   Fuzzy numbers

Special cases of fuzzy sets are fuzzy numbers, which are generalizations of our usual concept of numbers. While Boolean operations such as union, intersection or complement can be performed on general fuzzy sets, arithmetic operations such as addition or multiplication can only be performed on fuzzy sets defined on the set of real numbers. The simplest generalizations of the real numbers are the fuzzy numbers:

Definition 2.10 *A fuzzy subset A of the set of real numbers is called a fuzzy number if there is at least one z such that* $\mu_A(z) = 1$ *(normality* *assumption) and for every real numbers a, b, c with $a < c < b$*

$$\mu_A(c) \geq \min(\mu_A(a), \mu_A(b)) \qquad (2.2.1)$$

This second property is the so-called convexity assumption, meaning that the membership function of a fuzzy number usually consists of an increasing and decreasing part, and possibly flat parts. Figure 2.5 shows a convex and non-convex membership function on the real line.

It may be noted that the union or the intersection of fuzzy numbers are usually not themselves fuzzy numbers as it may be observed in Figures 2.2 and 2.3. The convexity assumption ensures that the h level sets and the support of a fuzzy number are intervals. In particular, the interval over which the membership function of a fuzzy number A is non-zero is called the support of A.

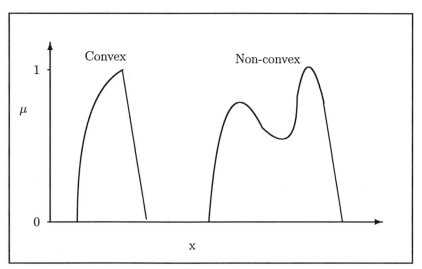

Figure 2.5. **A convex and non-convex membership function on the real line.**

Definition 2.11 *The support of the fuzzy number A is the set*

$$supp(A) = \{x \; ; \; \mu_A(x) > 0\}$$

The membership value of a real number reflects the "likeliness" of the occurrence of that number, the level sets (intervals in this case) reflect different sets of numbers with a given minimum likeliness (Kaufmann and Gupta, 1991). We use the word "likeliness" rather than "likelihood" to avoid semantic problems with both Bayesian and non-Bayesian probabilities. Zadeh (1978) suggested that $\mu_A(x)$ is the possibility that the parameter whose value is ill-known and described by the fuzzy number A takes value x. In this case $\mu_A(x) = 1$ means x is totally possible, and $\mu_A(x) = 0$ means x is impossible.

Any real number can be regarded as a fuzzy number with a single point support, and is called a "crisp number" in fuzzy mathematics. The simplest fuzzy numbers are the so-called triangular fuzzy numbers. The membership function of the triangular fuzzy number consists of an increasing and decreasing linear function — forming a triangle. The formal definition is:

Definition 2.12 *The fuzzy number $A = (a_1, a_2, a_3)_T$ with $a_1 \leq a_2 \leq a_3$ is a triangular fuzzy number if its membership function can be written in the form:*

$$\mu_A(x) = \begin{cases} 0 & \text{if } x \leq a_1 \\ \frac{x - a_1}{a_2 - a_1} & \text{if } a_1 < x \leq a_2 \\ \frac{a_3 - x}{a_3 - a_2} & \text{if } a_2 < x \leq a_3 \\ 0 & \text{if } a_3 < x \end{cases}$$

The support of the triangular fuzzy number $(a_1, a_2, a_3)_T$ is the intervall (a_1, a_3). Figure 2.6 shows the membership function of a triangular fuzzy number.

Example 2.3 *Consider the measurement of earthquakes by the so-called body wave technique. This technique as explained in Bárdossy et al. (1992a) essentially measures the amplitude of the quake as transmitted by the deep earth, rather than by the earth surface. It is known that the measuring instruments begin to saturate at about 7.00 (int. units), and that furthermore the measurements are by nature imprecise. It is thus natural to take the measured value as the peak of the membership function of a fuzzy set defined on the body wave scale of 1 to 9. If the measured value is far enough from the saturation zone, say 6, then a symmetric membership function assessed subjectively from an expert may be obtained; say the support is (5.8,6.2) as shown in Figure 2.7. On the other hand, a measurement of 7.1 is in the saturation zone, leading to a fuzzy set membership function that is skewed to the right with a support of, say (6.9,7.8), also shown in Figure 2.7. An analysis using fuzzy numbers is most useful in cases such as the latter, when the measurement errors are asymmetric and the assumptions of normality of errors is strongly violated.*

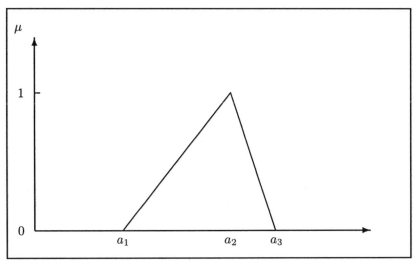

Figure 2.6. **Membership function of the triangular fuzzy number** $(a_1, a_2, a_3)_T$.

A special case of the triangular fuzzy numbers are the semi-infinite fuzzy numbers $(a_1, a_2, +\infty)_T$ defined as:

$$
\mu_A(x) = \begin{cases} 0 & \text{if } x \leq a_1 \\ \frac{x-a_1}{a_2-a_1} & \text{if } a_1 < x \leq a_2 \\ 1 & \text{if } a_2 < x \end{cases}
$$

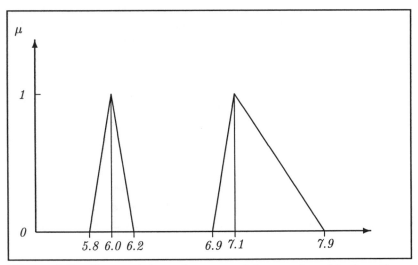

Figure 2.7. **Symmetric and asymmetric membership functions of measured earthquake magnitude.**

and $(-\infty, a_2, a_3)_T$:

$$\mu_A(x) = \begin{cases} 1 & \text{if } x \leq a_2 \\ \frac{a_3 - x}{a_3 - a_2} & \text{if } a_2 < x \leq a_3 \\ 0 & \text{if } a_3 < x \end{cases}$$

Other simple types of fuzzy numbers are the so-called trapezoidal fuzzy numbers, named after the form of their membership function:

Definition 2.13 *The fuzzy number $A = (a_1, a_2, a_3, a_4)_R$ with $a_1 \leq a_2 \leq a_3 \leq a_4$ is a trapezoidal fuzzy number if its membership function can be written in the form:*

$$\mu_A(x) = \begin{cases} 0 & \text{if } x \leq a_1 \\ \frac{x - a_1}{a_2 - a_1} & \text{if } a_1 \leq x \leq a_2 \\ 1 & \text{if } a_2 \leq x \leq a_3 \\ \frac{a_4 - x}{a_4 - a_3} & \text{if } a_3 < x \leq a_4 \\ 0 & \text{if } a_4 < x \end{cases}$$

The support of the trapezoidal fuzzy number $(a_1, a_2, a_3, a_4)_R$ is the interval (a_1, a_4). Figure 2.8 shows the membership function of a trapezoidal fuzzy number.

Example 2.4 *Let us define comfortable ambient relative humidity, given that the temperature is about $20^\circ C$ and the wind speed is low. We may say that under 20% the air is too dry and above 70%, too humid. Any value in the interval 35 to 50% is judged to be equally comfortable. The fuzzy set "comfortable relative humidity" may then be represented by the trapezoidal fuzzy number $K = (20, 35, 50, 75)_R$.*

Triangular fuzzy numbers are special cases of trapezoidal fuzzy numbers with $a_2 = a_3$. A pure interval is a particular case of a trapezoidal fuzzy number such that $a_1 = a_2$ and $a_3 = a_4 = b$, yielding $(a, a, b, b)_R$. Note that a triangular fuzzy number can only be reduced to a crisp number, not a pure interval.

Yet, other types of fuzzy numbers are the so called Left-Right or L-R fuzzy numbers (Dubois and Prade 1980a). In this case, the linear functions used in the definition of the triangular fuzzy numbers are replaced by monotonic functions:

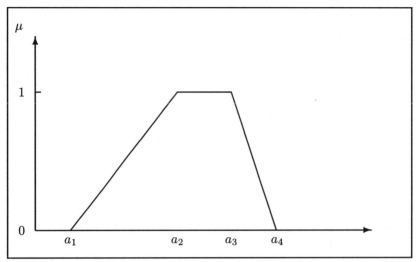

Figure 2.8. Membership function of the trapezoidal fuzzy number $(a_1, a_2, a_3, a_4)_R$.

Definition 2.14 *A fuzzy set $A = (a_1, a_2, a_3)$ on the set of real numbers is called an L-R fuzzy number if the membership of x can be calculated as:*

$$\mu_A(x) = \begin{cases} L(\frac{a_2 - x}{a_2 - a_1}) & \text{for } a_1 \leq x \leq a_2 \\ R(\frac{x - a_2}{a_3 - a_2}) & \text{for } a_2 \leq x \leq a_3 \\ 0 & \text{else} \end{cases}$$

Here L and R are continuous strictly decreasing functions defined on $[0, 1]$ with values in [0, 1] and satisfying the conditions:

$$L(x) = R(x) = 1 \text{ if } x \leq 0$$

$$L(x) = R(x) = 0 \text{ if } x \geq 1$$

The L-R fuzzy number A is denoted as:

$$A = (a_1, a_2, a_3)_{LR}$$

Triangular fuzzy numbers are special cases of L-R fuzzy number with $L(x) = R(x) = 1 - x$.

A reason for taking L-R fuzzy numbers is that they can sometimes express the transition between fulfillment and non-fulfillment better than the simple triangular fuzzy numbers since their shape can readily be controlled with the L and R functions. The linguistic hedges defined in Eqs. 2.1.3, 2.1.4, and 2.1.5 provide natural examples of L-R functions. Furthermore in some cases it is advantageous to have smooth, differentiable membership functions. Taking for example

$$L(x) = R(x) = \frac{1}{2}\left(\cos(\pi x) + 1\right) \qquad (2.2.2)$$

ensures that for any $a_1 < a_2 < a_3$ the resulting membership function is differentiable. Figure 2.9 shows the membership function of such a "bell shaped" L-R fuzzy number.

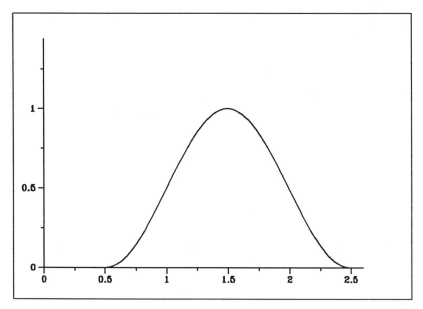

Figure 2.9. **Membership function of the L-R fuzzy number** $(0.5,1.5,2.5)_{LR}$ **with** $L(x) = R(x) = \frac{1}{2}\left(\cos(\pi x) + 1\right)$.

The shape of the membership function of a fuzzy number reflects how strong our belief is that we know the single "exact" value. The trapezoidal membership functions can also be generalized to L-R functions.

Definition 2.15 *A fuzzy set* $A = (a_1, a_2, a_3, a_4)$ *on the set of real numbers is called a trapezoidal L-R fuzzy number if the membership of x can* *be calculated as:*

$$\mu(x) = \begin{cases} L(\frac{a_2-x}{a_2-a_1}) & for \ \ a_1 \leq x \leq a_2 \\ 1 & for \ \ a_2 \leq x \leq a_3 \\ R(\frac{x-a_3}{a_4-a_3}) & for \ \ a_3 \leq x \leq a_4 \\ 0 & else \end{cases}$$

The L-R trapezoidal fuzzy number A is denoted as:

$$A = (a_1, a_2, a_3, a_4)_{LR}$$

Here again the shapes of the ascending and descending branches of the trapezoidal L-R number can be controlled. In particular membership functions derived from statistical data or a probabilistic analysis are by construction trapezoidal L-R fuzzy numbers. Such numbers are also widely used in fuzzy rule-base control.

Fuzzy numbers (or arbitrary fuzzy sets on the set of real numbers) can also be defined with the help of piecewise linear membership functions. For this a set of breakpoints $x_0 < x_1 < \ldots < x_L$ and the corresponding membership values $\mu(x_0), \mu(x_1), \ldots < \mu(x_L)$ have to be given, and it is assumed that the membership function is linear between these points. The advantage of this formulation is that most continuous functions can be closely approximated by piecewise linear functions, and the operations on the latter are usually simple.

One can also define fuzzy vectors in an n-dimensional space. For this purpose the Cartesian product is introduced:

Definition 2.16 *If $A_1, \ldots, A_I$ are fuzzy sets in $X_1, \ldots, X_I$ then the Cartesian product of $A_1, \ldots, A_I$, that is $A_1 \times \ldots \times A_I$ is defined by the membership function:*

$$\mu(x_1, \ldots, x_I) = \min(\mu_{A_1}(x_1), \ldots, \mu_{A_I}(x_I)) \qquad x_i \in X_i, i = 1, \ldots, I. \tag{2.2.3}$$

By definition the h level set of the Cartesian product is the same as the traditional Cartesian product of the h level sets. Formally:

$$(A_1 \times \ldots \times A_I)(h) = A_1(h) \times \ldots \times A_I(h)$$

This could also have been the definition of the Cartesian product.

2.2.1 Operations on fuzzy numbers

As fuzzy numbers are generalizations of real numbers, it is desirable to define the usual (arithmetic) operations (addition, multiplication, etc.) on them. Furthermore, the newly defined operations should reduce to the usual ones if applied to crisp numbers. The *extension principle* (Zadeh 1965) is a method of extending point to point operations to fuzzy sets. It is the basic tool for the development of fuzzy arithmetic.

Definition 2.17 *If X and Y are two sets, and f is a point to point mapping from X to Y*

$$f : X \longrightarrow Y \quad for\ every \quad x \in X \quad f(x) = y \in Y$$

then f can be extended to operate on fuzzy subsets of X in the following way:
Let A be a fuzzy subset of X with membership function μ_A, then the image of A in Y is the fuzzy subset B with the membership function

$$\mu_B(y) = \begin{cases} \sup\{\mu_A(x); y = f(x), x \in X\} \\ 0 \quad if\ there\ is\ no\ x \in X \quad such\ that\ f(x) = y \end{cases} \tag{2.2.4}$$

To illustrate the above definition, consider the following numerical example.

Example 2.5 *Suppose $f(x)$ maps all three elements x_1, x_2 and x_3 to the same element y:*

$$f(x_1) = f(x_2) = f(x_3) = y$$

If $A = \{(x_1, 0.5), (x_2, 0.2), (x_3, 0.7)\}$ then $B = \{(y, \max(0.5, 0.2, 0.7))\} = \{(y, 0.7)\}$. Figure 2.10 illustrates this example.

From the extension principle it follows that for every h-level set $A(h)$, the transformed h-level set $B(h)$ is given by

$$B(h) = f(A(h)) = \{y; y = f(x), x \in A(h)\}$$

which is the definition of an ordinary function defined on $A(h)$. The extension principle combined with the Cartesian product (Def. 2.16) is used to define operations on fuzzy numbers. Let $\odot$ be a bivariate

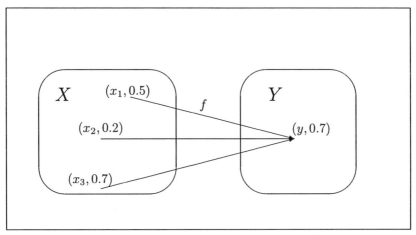

Figure 2.10.   The extension principle.

arithmetic operation (for example addition or multiplication) defined on
two fuzzy numbers A and B. Then

$$C = A \odot B$$

is the fuzzy number with the membership function defined as:

$$\mu(y) = \sup\{\min(\mu_A(x_1), \mu_B(x_2)) \; ; \; \text{such that } y = x_1 \odot x_2\}$$

Example 2.6 *First consider the application of the extension formula to
the univariate function* $\log(x)$. *The image of the triangular fuzzy number*
$A = (2,3,4)_T \log(A)$ *is to be found using the extension principle. As*
$\log(x)$ *is a monotonic function for each real number* y, *there is a single
value* x *for which* $\log(x) = y$. *Thus, the extension principle yields* $B =
\log(A)$ *with the membership function:*

$$\mu_B(y) = \begin{cases} e^y - 2 & \text{if} \quad \log(2) \le y < \log(3) \\ 4 - e^y & \text{if} \quad \log(3) \le y \le \log(4) \end{cases} \tag{2.2.5}$$

As a bivariate example the addition of two fuzzy numbers A *and* B *is
considered. The h-level set of the Cartesian product* C *of the two numbers*

is by definition:

$$C(h) = A(h) \times B(h)$$

The function to be extended is:

$$f(x, y) = x + y$$

Thus for the result $D = A + B$ the h-level set is obtained by

$$D(h) = A(h) + B(h)$$

Performing this for each h level one has the membership function of the sum. The addition can be simplified for triangular or L-R fuzzy numbers. For example:

$$(a_1, a_2, a_3)_{LR} + (b_1, b_2, b_3)_{LR} = (a_1 + b_1, a_2 + b_2, a_3 + b_3)_{LR}$$

Assuming the L and R functions corresponding to A and B are the same

2.2.2   Fuzzy mean and median

As it will be shown, once a fuzzy rule has been applied, it is often necessary to "defuzzify" a fuzzy set — i.e., to replace it with a single crisp element representative of the set, for example the crisp class of a fuzzy object. A natural way to do this is to take the element with the highest membership value. The problem with this procedure applied over a continuous domain is that in many cases the maximum is not unique: thus, in the case of a trapezoidal fuzzy number $(a_1, a_2, a_3, a_4)_R$ all number of the interval $[a_2, a_3]$ have the maximal membership value of one. Furthermore the maximal membership may correspond to independent variable values which are not really representative of the fuzzy set. For example the triangular fuzzy numbers $(0, 3, 3)_T$ and $(3, 3, 9)_T$ both have 3 as the element with the highest membership value but the crisp "location" of the two fuzzy numbers is subjectively felt as being different.

In the case of fuzzy sets defined on the set of real numbers the "location" of a fuzzy set can be described by the fuzzy mean M (also called center of gravity or centroid). This is the number for which the part of the membership function on the left of this number is in equilibrium with the right side. The equilibrium occurs when the moments corresponding to the two sides equal. This yields:

Definition 2.18 *The fuzzy mean of the fuzzy set A defined on the real line is the number $M(A)$ for which:*

$$\int_{-\infty}^{M(A)} (M(A) - t)\mu_A(t)\, dt = \int_{M(A)}^{+\infty} (t - M(A))\mu_A(t)\, dt \qquad (2.2.6)$$

Note that not all fuzzy numbers have a fuzzy mean: for example, $(a_1, a_2, +\infty)_T$ has no finite fuzzy mean. The calculation of the fuzzy mean can be simplified by reworking Eq. (2.2.6) as:

$$M(A) \int_{-\infty}^{M(A)} \mu_A(t)\, dt - \int_{-\infty}^{M(A)} t\mu_A(t)\, dt =$$

$$= \int_{M(A)}^{+\infty} t\mu_A(t)\, dt - M(A) \int_{M(A)}^{+\infty} \mu_A(t)\, dt$$

from which one has

$$M(A) \int_{-\infty}^{+\infty} \mu_A(t)\, dt = \int_{-\infty}^{+\infty} t\mu_A(t)\, dt$$

So $M(A)$ can directly be calculated using:

$$M(A) = \frac{\int_{-\infty}^{+\infty} t\mu_A(t)\, dt}{\int_{-\infty}^{+\infty} \mu_A(t)\, dt} \qquad (2.2.7)$$

If $A = (a_1, a_2, a_3)_T$, then it can be shown that

$$M(A) = \frac{a_1 + a_2 + a_3}{3} \qquad (2.2.8)$$

For more general fuzzy sets the fuzzy mean can be calculated using the following proposition:

Proposition 2.1 *Consider a fuzzy set A with a piecewise linear membership function having breakpoints: $x_0 < x_1 < \ldots < x_L$. The fuzzy mean of A can then be calculated as:*

$$M(A) = \frac{\sum_{l=0}^{L-1}(x_{l+1}-x_l)\left(\frac{2x_{l+1}+x_l}{6}\mu(x_{l+1}) + \frac{x_{l+1}+2x_l}{6}\mu(x_l)\right)}{\sum_{l=0}^{L}(x_{l+1}-x_l)\frac{\mu(x_{l+1})+\mu(x_l)}{2}} \quad (2.2.9)$$

As each continuous function can be closely approximated by a piecewise linear function, the above formula is of very general use.

The fuzzy mean of the L-R fuzzy number $(a_1, a_2, a_3)_{LR}$ can also be calculated directly as:

$$\int_{-\infty}^{+\infty} t\mu_A(t)\,dt = \int_{a_1}^{a_2} tL\left(\frac{a_2-t}{a_2-a_1}\right)dt + \int_{a_2}^{a_3} tR\left(\frac{t-a_2}{a_3-a_2}\right)dt \quad (2.2.10)$$

After some algebra one gets:

$$M(A) = \frac{a_2(a_2-a_1)L^* - (a_2-a_1)^2 L^{**} + a_2(a_3-a_2)R^* + (a_3-a_2)^2 R^{**}}{(a_2-a_1)L^* + (a_3-a_2)R^*}$$

$$(2.2.11)$$

with

$$L^* = \int_0^1 L(t)\,dt$$

$$L^{**} = \int_0^1 tL(t)\,dt$$

$$R^* = \int_0^1 R(t)\,dt$$

$$R^{**} = \int_0^1 tR(t)\,dt$$

Note that these starred quantities only depend on the L, R functions and not on (a_1, a_2, a_3).

The fuzzy mean is a good location parameter — small changes of the membership function only cause small changes in the mean. This fact is formalized in the following proposition:

Proposition 2.2 *The fuzzy mean $M(A)$ is continuous for fuzzy sets with the same bounded support $[-K, K]$: For each $\varepsilon > 0$ there is a $\delta > 0$ such that for two fuzzy sets A and B*

$$|M(A) - M(B)| < \varepsilon$$

holds if

$$|\mu_A(x) - \mu_B(x)| < \delta$$

for all $x \in [-K, K]$

The proof of this proposition is given in the Appendix. Another "location" type parameter of a fuzzy number is its median. The median cuts the membership function into two equal area parts. (These two do not have to be in a "physical" equilibrium as in the case of the fuzzy mean.) The formal definition of the fuzzy median is:

Definition 2.19 *The median $m(A)$ of fuzzy set A is defined by the equation*

$$\int\limits_{-\infty}^{m(A)} \mu_A(t)\, dt = \int\limits_{m(A)}^{+\infty} \mu_A(t)\, dt \qquad (2.2.12)$$

Note that, in contrast with mean $M(A)$, the area under the membership function does not appear as a normalizing factor. If $A = (a_1, a_2, a_3)_T$ then it can be shown that

$$m(A) = a_1 + \left[\frac{(a_2 - a_1)(a_3 - a_1)}{2}\right]^{\frac{1}{2}} \quad \text{if } a_2 > \frac{a_1 + a_3}{2} \quad \text{and,}$$

$$m(A) = a_3 + \left[\frac{(a_3 - a_1)(a_3 - a_2)}{2}\right]^{\frac{1}{2}} \quad \text{if } \frac{a_1 + a_3}{2} > a_2 \quad (2.2.13)$$

For right angle TFN's

$$\text{if } a_1 = a_2, \quad m(A) = a_3 - \frac{(a_3 - a_1)}{\sqrt{2}}$$

$$\text{if } a_2 = a_3, \quad m(A) = a_1 + \frac{(a_3 - a_1)}{\sqrt{2}} \qquad (2.2.14)$$

The calculation of the median for a piecewise linear membership function is slightly more difficult than the calculation of the fuzzy mean.

The median is a good "location" type measure for fuzzy numbers; however, for an arbitrary fuzzy set defined on the set of real numbers, difficulties may arise as shown in the next example.

Example 2.7 Let $A = (1, 2, 3)_T$ and $B = (4 + \epsilon, 5, 6 - \epsilon)_T$ with $|\epsilon| < 1$, and let $C = A \cup B$. As the support of the two fuzzy numbers is disjoint, the membership function of C may be written as:

$$\mu_C(x) = \max(\mu_A(x), \mu_B(x)) = \mu_A(x) + \mu_B(x)$$

For A and B one has:

$$\int_{-\infty}^{+\infty} \mu_A(t)\, dt = 1 \qquad \int_{-\infty}^{+\infty} \mu_B(t)\, dt = 1 - \epsilon$$

So for $\epsilon > 0$ the median is such that

$$\int_{-\infty}^{m(C)} \mu_A(t)\, dt = 1 - \frac{\epsilon}{2}$$

Thus for $m(C)$ one has

$$\frac{(3 - m(C))^2}{2} = \frac{1}{2}(1 + 1 - \epsilon) = \frac{\epsilon}{2}$$

from which it follows that $m(C) = 3 - \sqrt{\epsilon}$. Similar calculations yield:

$$m(C) = \begin{cases} 4 - \epsilon + \sqrt{\epsilon(1 + \epsilon)} & \text{if } 0 > \epsilon > -1 \\ \text{undefined in } [3, 4] & \text{if } \epsilon = 0 \\ 3 - \sqrt{\epsilon} & \text{if } 1 > \epsilon > 0 \end{cases}$$

This relation means that the median is not continuous, a slight difference in the membership function can cause a jump in the median value. In *contrast, the fuzzy mean, as per Proposition 2.2, is a continuous function of the disturbance. Specifically:*

$$\frac{\int\limits_{-\infty}^{+\infty} t\mu_A(t)\, dt + \int\limits_{-\infty}^{+\infty} t\mu_B(t)\, dt}{\int\limits_{-\infty}^{+\infty} \mu_A(t)\, dt + \int\limits_{-\infty}^{+\infty} \mu_B(t)\, dt} = \frac{1 M(A) + (1-\epsilon)M(B)}{2-\epsilon} \frac{1\cdot 2 + (1-\epsilon)\cdot 5}{2-\epsilon}$$

Thus one finds

$$M(C) = \frac{7 - 5\epsilon}{2 - \epsilon}$$

which is a continuous function of ϵ.

It has to be noted that if the membership function μ_A of a fuzzy set A (defined on the real numbers) is continuous and the support of A is a finite interval, then the median is also continuous, i.e., small changes of the membership function cause only small changes of the median.

On the other hand, the median has the advantage that it can be defined for fuzzy sets on discrete ordered sets X, like the set of natural numbers. In this case the median $m(A)$ can be defined as the value

$$\left| \sum_{x \le m(A)} \mu_A(x) - \frac{\mathrm{car}(A)}{2} \right| \quad \text{minimal} \qquad (2.2.15)$$

This means that the median is the element that best divides the fuzzy set into two subsets with equal cardinality (considering the order as well). This definition of the median is not always unique.

2.2.3   Distance between fuzzy numbers

Besides the arithmetic defined on fuzzy numbers a distance measure is also often needed (to express similarity, for optimization etc.). A distance between fuzzy numbers should be defined so that its restriction to the crisp numbers should yield the usual Euclidean distance. Diamond (1988) suggested to use $d(A, B)$ defined as:

$$d^2(A, B) = \left((a_1 - b_1)^2 + (a_2 - b_2)^2 + (a_3 - b_3)^2 \right) \qquad (2.2.16)$$

as the distance between the two triangular fuzzy numbers $A = (a_1, a_2, a_3)_T$ and $B = (b_1, b_2, b_3)_T$. This is in fact a distance and it is a very simple generalization of the Euclidean one. However, it is only defined on triangular fuzzy numbers and cannot be extended to arbitrary L-R fuzzy numbers. Furthermore this distance takes differences at the 0 membership level as much into account as differences at the 1 level. However we believe that differences at the 1 membership level are more important than those at or near the 0 membership level because they are subject to smaller uncertainty. To deal with this problem a weighting function can be introduced. This leads to the following definition of a distance defined for all L-R fuzzy numbers.

Definition 2.20 *Let the Hagaman squared distance between fuzzy numbers $A = (a_1, a_2, a_3)_{LR}$ and $B = (b_1, b_2, b_3)_{LR}$, with L_A, L_B, R_A and R_B being the L-R functions of A and B be defined as:*

$$D^2(A, B, f) = \int_0^1 \left\{ \left(a_2 - (a_2 - a_1)L_A^{-1}(q) - b_2 + (b_2 - b_1)L_B^{-1}(q) \right)^2 + \right.$$

$$\left. \left(a_2 + (a_3 - a_2)R_A^{-1}(q) - b_2 - (b_3 - b_2)R_B^{-1}(q) \right)^2 \right\} f(q)\, dq \quad (2.2.17)$$

Here f is a continuous function defined on $[0, 1]$ with the following properties:

$$f(q) > 0 \text{ if } q > 0$$

$$\int_0^1 f(q)\, dq = \frac{1}{2}$$

The function f serves as a weighting function. The distance is a weighted sum (integral) of the differences between the two numbers at selected levels.

It is reasonable to assume that $f(q)$ is an increasing function. This means that the distance between two fuzzy numbers is greater if they differ at high membership levels than in the case of differences at low membership levels.

It can be proved (Bárdossy et al., 1992a) that

$$D(A, B, f) = \sqrt{D^2(A, B, f)}$$

is a distance on the set of all L-R fuzzy numbers.

The calculation of the distance can be simplified using the following quantities:

$$\int_0^1 L_A^{-1}(q) f(q)\, dq = L_A^*$$

$$\int_0^1 R_A^{-1}(q) f(q)\, dq = R_A^*$$

$$\int_0^1 (L_A^{-1}(q))^n f(q)\, dq = L_A^{n*}$$

$$\int_0^1 (R_A^{-1}(q))^n f(q)\, dq = R_A^{n*}$$

$$\int_0^1 L_A^{-1}(q) L_B^{-1}(q) f(q)\, dq = L_{AB}^{**}$$

$$\int_0^1 R_A^{-1}(q) R_B^{-1}(q) f(q)\, dq = R_{AB}^{**}$$

$$D^2(A, B, f) = (a_2 - b_2)^2 + 2(a_2 - b_2)$$

$$((a_3 - a_2) R_A^* + (b_2 - b_1) L_B^* - (b_3 - b_2) R_B^* - (a_2 - a_1) L_A^*) +$$

$$+ (a_2 - a_1)^2 L_A^{2*} - 2(a_2 - a_1)(b_2 - b_1) L_{AB}^{**} + (b_2 - b_1)^2 L_B^{2*} +$$

$$+ (a_3 - a_2)^2 R_A^{2*} - 2(a_3 - a_2)(b_3 - b_2) R_{AB}^{**} + (b_3 - b_2)^2 R_B^{2*} \quad (2.2.18)$$

This equation gives a computationally easier form than the originally defined distance. The distance between fuzzy and crisp numbers can also be calculated using Eq. 2.2.18. For example in the case of triangular fuzzy numbers:

$$L(q) = R(q) = 1 - q$$

and supposing the weighting function is linear ($f(q) = q$) then:

$$L_A^* = R_A^* = \frac{1}{6}$$

$$L_A^{2*} = R_A^{2*} = L_{AB}^{**} = R_{AB}^{**} = \frac{1}{12}$$

and the Hagaman squared distance can be written as:

$$D^2(A, B, f) = (a_2 - b_2)^2 + \frac{1}{3}(a_2 - b_2)(a_3 - 2a_2 + a_1 - b_3 + 2b_2 - b_1) +$$

$$+ \frac{1}{12}\left((a_2 - a_1 - b_2 + b_1)^2 + (a_3 - a_2 - b_3 + b_2)^2\right) \qquad (2.2.19)$$

To illustrate the difference between th distances, consider the following example:

Example 2.8 *Consider three fuzzy numbers:*

$$A_1 = (2, 2, 2)_T$$

$$A_2 = (1, 2, 3)_T$$

and for $0 < \varepsilon < 1$ let

$$c = \left(2 + \sqrt{\frac{2}{3}(1 - \varepsilon^2)}\right)$$

and

$$A_3 = (c - \varepsilon, c, c + \varepsilon)_{LR}$$

with linear L-R functions and let $f(q) = q$ In this case one has:

$$D^2(A_1, A_2, f) = \frac{1}{6}$$

$$D^2(A_1, A_3, f) = \frac{2}{3}(1 - \varepsilon^2) + \frac{1}{6}\varepsilon^2$$

The Diamond distance d is found to be

$$d^2(A_1, A_2) = d^2(A_1, A_3) = \frac{2}{3}$$

Figure 2.11 shows the configuration of the three numbers. Note that using

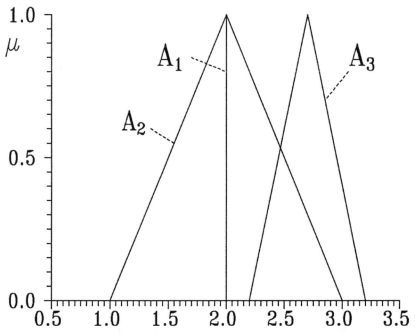

Figure 2.11.   Fuzzy numbers A_1, A_2 and A_3 for Example 2.8.

the distance D, A_1 is closer to A_2 than to A_3 which is reasonable. The distance d gives the intuitively incorrect result of equal distances.

This distance can also be used to find the closest crisp number to a given fuzzy number. This is another "location" type measure like the fuzzy mean or the median. For a triangular fuzzy number $A = (a_1, a_2, a_3)_T$ the closest crisp number using weighting function $f(q) = q$ is the number a which minimizes:

$$D^2(A, a, f) = (a_2 - a)^2 + \frac{1}{3}(a_2 - a)(a_3 - 2a_2 + a_1) +$$

$$+ \frac{1}{12}\left((a_2 - a_1)^2 + (a_3 - a_2)^2\right) \qquad (2.2.20)$$

To find the minimum the derivative of the above function with respect to a has to be 0, namely:

$$-2(a_2 - a)^2 - \frac{1}{3}(a_3 - 2a_2 + a_1) = 0$$

from which the closest crisp number a is:

$$a = \frac{1}{6}a_1 + \frac{2}{3}a_2 + \frac{1}{6}a_3 \qquad (2.2.21)$$

Similar calculations show that the closest crisp number for the distance d for triangular fuzzy numbers defined in Equation (2.2.16) is:

$$a = \frac{1}{3}a_1 + \frac{1}{3}a_2 + \frac{1}{3}a_3 \qquad (2.2.22)$$

which is the fuzzy mean. These distance based crisp numbers are reasonable location parameters, but in contrast to the fuzzy mean and the median there is no simple generalization that could be used for any fuzzy sets on the real line.

2.3   Assessment of the membership functions

A crucial point in applying fuzzy methods is the assessment of the membership functions. There are only a few methods published in the fuzzy literature that give advice in doing this (Civanlar and Trussel, 1986; Dubois and Prade, 1986; Türksen, 1991). Further membership experiments can be found in Hisdal (1994). A very simple way of defining a fuzzy number A with respect to a parameter x is by assessing three numbers:

1. the most credible value x^* — assigned a membership value of 1.

2. the number x^- which is almost certainly exceeded by the parameter value — assigned a membership value of 0.

3. the number x^+ which is almost certainly not exceeded by the parameter value — also assigned a membership value of 0.

Let the membership function be defined as 0 outside the interval (x^-, x^+) of possible values (or support), and taken to be piecewise linear in between: the triangular fuzzy number $A_T = (x^-, x^*, x^+)_T$ has thus been constructed. Note that the resulting membership function is not necessarily symmetrical. This result is different from the usual assumption of normally or at least symmetrically distributed error around the most credible value.

Other techniques are available to assess membership functions depending on the type of imprecision described by a given fuzzy set. As pointed out in Duckstein et al. (1989), a fuzzy set membership function may represent a value function, for example harshness of winter with respect to wildlife population; in this case techniques for assessing value functions may be used (Duckstein and Heidel, 1988). In other cases a fuzzy set membership function may be analogous to a prior probability, for example in the case of subjective forecasting of a hydrometeorological variable; then a technique for assessing prior probability may be used (Berger 1985). Türksen (1991) presents further general techniques for assessing membership functions.

As membership functions are often related to the perception by humans, it might be reasonable to take the human response to outside stimuli into account. In a recent study, Ge and Laurig (1993) investigated the human perception of weight and assessed membership functions using this information. Numerous experiments show that the psychological response to stimuli increases exponentially with the increase of the stimulus. This relationship can be written in the form (Stevens 1975):

$$R = cS^p \qquad (2.3.1)$$

where c and p are constants depending on the properties of the stimuli. This means that if a membership function is thought to describe, for example, a weight to be lifted, then a triangular membership function defined on the perception scale becomes a usually non symmetric L-R fuzzy number on the weight scale (if $p \neq 1$). Ge and Laurig (1993) found that the appropriate shape of a membership function can be obtained using an exponential transformation. Figure 2.12 shows a characteristic membership function for weights.

Once a membership function has been assessed, a sensitivity analysis may be performed to find out if further refinement is necessary. If it is found that the model behavior is sensitive to the shape or support of the membership functions then it is possible to use neural nets to improve an initial assessment of these functions (Bardossy et al., 1993c).

2.4   Fuzzy sets, possibilities and probabilities

Fuzzy sets appear to be in many respects similar to probability distributions defined on the same set or domain. The use of the unit interval in both cases and the fact that both concepts describe uncertainty make this analogy even more visible. However, this apparent analogy is quite superficial, for there are many essential differences (a few that are of in-

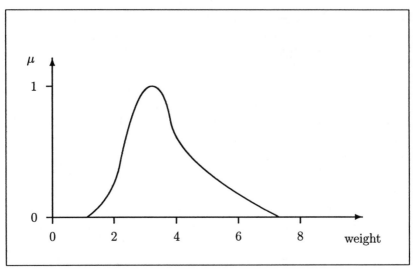

Figure 2.12. A characteristic membership function for weight perception.

terest in this monograph are listed below). Further details may be found in a special issue of *IEEE Transactions on Fuzzy Systems* (Dubois and Prade, 1994a; Klir 1994; Kosko 1994), which is in fact strongly recommended reading for anyone interested in this topic.

2.4.1   Possibility

In a probabilistic representation of uncertainty, the additivity is the basic assumption. However, additivity is not always appropriate as for example in the case of ignorance. As an alternative to probability, possibility theory has been suggested by Zadeh (1978). The definition of a possibility measure Π is similar to that of a probability measure, with the difference that the possibility of a sum is the supremum of the individual possibilities. The exact definition is:

Definition 2.21 *A possibility measure on the set X is the function Π assigning values to subsets of X with the properties:*

1. $\Pi(\emptyset) = 0$ and $\Pi(X) = 1$

2. If $A \subset B$ then $\Pi(A) \leq \Pi(B)$

3. $\Pi(\bigcup_{i \in I} A_i) = \sup_{i \in I} \Pi(A_i)$ for any index set I

Using this definition the possibility of a set can be calculated as

$$\Pi(A) = \sup_{x \in A} \Pi(\{x\})$$

Taking

$$\Pi(\{x\}) = \mu(x)$$

one can relate a possibility measure and a fuzzy set. This relationship is similar to the one between probability measure and density. However note that only normal membership functions ($\sup_x \mu(x) = 1$) define possibility measures. More details on possibility theory can be found in Dubois and Prade (1988).

2.4.2 Membership functions versus probabilities

Probability is defined on a measure space, whereas fuzzy sets do not require such an additional structure on the universe X.

In probability theory an event involves only crisp elements and is thus defined precisely; for example, "the height of a certain person is less than 175 cm" is defined as precisely as one wishes to measure height. The uncertainty bears on whether an element belongs to the set or not.

In the case of fuzzy sets the element that takes the place of an "event" is usually well defined; for example, John is 175 cm. So, there is no uncertainty about John's size. The uncertainty is whether or not (and to what extent) that well-defined element fulfills the definition of the set. In this example, does John belong to the set of tall individuals? Dubois and Prade (1993) provide a different way to look at the membership value $\mu(175)$, namely, as the degree of similarity between 175 cm and the prototypes of tall. Either way of considering a membership function value leads to the same fuzzy rule-based model.

Still another question occurs if we only know "John is tall" and ask if John measures 175cm. Then $\mu_T(175)$ is the degree of possibility that John is tall (Dubois and Prade, 1988).

For a random variable ξ the probability distribution function contains interval type information $F(x) = P(\xi < x)$. In contrast, a membership function value contains information concerning the individual element, namely, the likeliness or degree of credibility that it fulfills the definition of the fuzzy set.

There is a certain formal similarity between a fuzzy number membership function and the density function of a random variable. However, the density function is normalized by the area under the function, whereas the fuzzy number is in contrast normalized by its maximal value. Furthermore, probability density functions can be bimodal which is not allowed for fuzzy numbers because of the convexity assumption. The most important difference lies in the operations: the density function of the sum of two random variables is calculated with some type of convolution integral (determined by the dependence between the random variables). The sum of two fuzzy numbers is calculated with the extension principle which involves no sum-product operation.

In real life cases, both kinds of uncertainty may occur simultaneously leading to the concept of fuzzy probability. A combination of the concepts is also possible and sometimes even necessary, as illustrated by cases of fuzzy failure (Bárdossy and Bogárdi, 1989; Shrestha and Duckstein, 1993). More formally, Zadeh (1973) has enounced the so-called consistency principle, stating that, over any given interval, the possibility of a fuzzy event, related to the membership function, should be larger than its probability over any interval of X. Civanlar and Trussel (1986) use this principle for a rigorous construction of membership functions from statistical data: the procedure they develop, however, involves fairly complex optimization and numerical schemes. It is our experience that a probability density function defined on (a_1, a_4) may be transformed in an ad hoc way into a trapezoidal L-R number $(a_1, a_2, a_3, a_4)_{LR}$ as follows so as to follow approximately at least the consistency principle. Let d_L and d_R be a measure of dispersion of a probability density on the left and right of the mode m_0. If M is the mean and σ, the standard deviation of the density, then one may pose for these measures of deviation.

In the case $m_0 \geq M$

$$d_L = \sigma + m_0 - M$$
$$d_R = \sigma$$

The 1-level set is taken as $(a_2, a_3) = (m_0 - d_L, m_0 + d_R)$. The left branch of the membership function (a_1, a_2) is affine to the left side of the probability density over the same interval; the right branch (a_3, a_4) is constructed analogously from the right side of the density over the interval (a_3, a_4). In any case, a sensitivity analysis of the results over the shape of the membership functions of the premises and consequences is always in order.

In the case $m_0 < M$, one may then pose:

$$d_L = \sigma$$

$$d_R = \sigma - m_0 + M$$

This second case is illustrated in Fig. 2.13. A bibliography on the passage from possibility to probability, together with techniques for doing this, may be found in Dubois and Prade (1993).

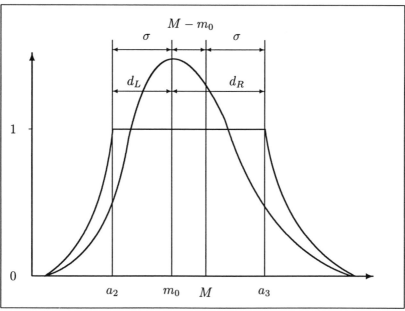

Figure 2.13.   Transformation of a probability density $f(x)$ into a fuzzy set membership function $\mu(x)$, in the case $m_0 < M$.
M = mean of $f(x)$.
σ = standard deviation of f(x)
m_0 = mode of $f(x)$

3

Fuzzy rules

As pointed out in the Introduction, rules are used very widely both for descriptive and control purposes. However, sometimes it is quite difficult to formulate these rules in precise (computer applicable) form. Quite often one faces the problem of imprecise information — partly due to the imprecise measurements and partly due to the non existence of suitable measures. For example, there is no accepted measure for headache or for severity of an extreme drought (Shrestha e. al., 1994; Pesti et al., 1994). A further problem in formulating rules arises from the fact that exceptions cannot be tolerated by classical true/not true type of rules.

The Introduction briefly illustrated how fuzzy rules may be used whenever an explicit function is impractical or difficult to define, derive or calibrate.

Assuming that a function defined on a set X with values in set Y exists, then the crisp rule is that for every element $x \in X$ there should be one and only one value $y \in Y$. The medical decision aid example developed in Blinowska et al. (1992) and described in Chapter 10 shows that for one given patient file with experimental data — which may be considered as a point x in an N dimensional space — each of 5 specialists may come up with a different vector of answers to 5 questions: this vector may have been the value y of the function at point x. In that example, three specialists may reach contradictory conclusions: two say the risk is mild, one says the risk is severe. A crisp rule-based system would conclude that there is no rule, since it tolerates no exception and no error, while a fuzzy rule-based system will still use this information in the model calibration phase, since it tolerates exceptions. Not only wide differences in opinion but also difficulties with measurement errors may thus be overcome. The medical diagnosis example described in Zhang and Duckstein (1993) leads to similar remarks.

In different words, Blinowska et al. (1992) could have assumed a relationship for the risk y as a function of hypertension x, of the form:

$$y = a + bx \tag{3.0.1}$$

43

Then some criterion, such as least squares, would have been used to estimate coefficients a and b.

Alternatively, a and b may be considered as fuzzy numbers $\hat{a}$ $\hat{b}$, and determined by some optimization procedure (Bárdossy 1990, Bárdossy et al. 1990, 1992a). Still, a given functional form must be assumed for both the fuzzy regression and model results (Bárdossy et al., 1993b).

A rule consists of arguments coupled by logical operators forming a logical expression and a corresponding consequence. If the conditions of the rule are fulfilled (the logical expression is true) then the consequence has to be true. The logical expression is usually formulated with simple uni- and bivariate logical operators.

In Boolean logic, two values 0 (false) and 1 (true) can be assigned to any statement or rule. Table 3.1 where XOR stands for either ..., or,..., shows the truth values of different logical operators. The same operators will be used initially, to define our fuzzy rules.

Table 3.1.   The Boolean truth value of basic logical operators

A	B	NOT(A)	A AND B	A OR B	A XOR B	A $\Rightarrow$ B
1	1	0	1	1	0	1
1	0	0	0	1	1	0
0	1	1	0	1	1	1
0	0	1	0	0	0	1

The implication operator A $\Rightarrow$ B is also shown in the table; however, it will require special treatment in the fuzzy case.

With a fuzzy rule, no explicit functional form is assumed, binary logic is replaced by fuzzy logic where a statement and its opposite may both be "true" to a certain — hopefully different degree. For example "severe" and "moderate" pathology may both be "true" for a given patient. As for fuzzy A and B the "truth value" can vary between 0 and 1. The Boolean table has to be extended to cope with such situations in a plausible manner.

3.1   The structure of a fuzzy rule

In this section, rules with fuzzy arguments are defined. Then, the Boolean "truth table" 3.1 is extended to the case of fuzzy arguments. This way a "truth value" can be assigned not only to the simple statements used for the definition of fuzzy sets of the type: a is A, but also to complex logical coupling of such simple statements.

3.1.1 The form of a fuzzy rule

A fuzzy rule consists of a set of arguments $A_{i,k}$ in the form of fuzzy sets with membership functions $\mu_{A_{i,k}}$ and a consequence B_i also in the form of a fuzzy set.

$$\text{If } a_1 \text{ is } A_{i,1} \odot a_2 \text{ is } A_{i,2} \odot \ldots \odot a_K \text{ is } A_{i,K} \text{ then } B_i \qquad (3.1.1)$$

Note that in general the order in which a statement is evaluated plays a central role. The operator $\odot$ stands here for AND or OR or XOR. The statements "a_k is $A_{i,k}$" is, for simplicity sake, replaced by $A_{i,k}$ in the following text. The rule (Eq. 3.1.1) will thus be written as

$$\text{If } A_{i,1} \odot A_{i,2} \odot \ldots \odot A_{i,K} \text{ then } B_i \qquad (3.1.2)$$

Verbal rules are often translated into fuzzy rules using linguistic variables.

Example 3.1 *If it is cold and I have a long way to walk, then I usually take my coat.*
Here the fuzzy set $A_{1,1}$ represents the temperature. "Cold" might be characterized with a fuzzy set with membership 1 for $-10°C \leq T \leq 0°C$, 0 for $T \geq 15°C$, and $T \leq -20°C$ and linear in between. This is the trapezoidal fuzzy number $(-20, -10, 0, 15)_R$. The fuzzy set "long walk" $A_{1,2}$ can also be characterized by a fuzzy number $(200, 1500, 4000)_T$. meters.

Figure 3.1 shows a fuzzy rule in a graphical form, A_{i1} and A_{i2} being the fuzzy rule arguments and B_i the corresponding fuzzy response.
Instead of the usual Boolean case when a rule can either be applied with certainty or not, here a partial applicability is also possible. There may be cases where a few different rules with different consequences can to a certain degree be applied to the same premises. However rules are sought to provide responses — therefore, methods to find responses reflecting the applicability of the rules have to be developed. To achieve this goal, the first step is to find a truth level for the applicability of a fuzzy rule.

3.1.2 Implication operators

A possibility to consider rules is to use the implication operations: "If X is A THEN Y is B". The fuzzy sets A and B are defined on the sets U and V, respectively. The possibility distribution on the resulting set

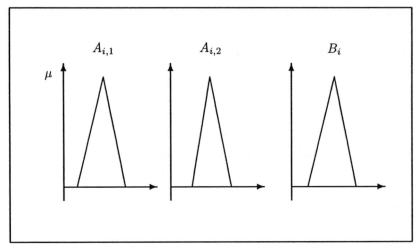

Figure 3.1.   A fuzzy rule.

V can be calculated after Dubois and Prade (1988) with the help of the conditional possibility distribution $\pi_{Y|X}$. The formula for any given y is:

$$\pi_Y(v) = \sup_u \{t(\pi_{Y|X}(v, u), \pi_X(u))\} \qquad (3.1.3)$$

with t being a triangular norm as defined in Definition 2.7. In this case, using the membership functions μ_A and μ_B as possibility distributions one obtains the inequality:

$$\mu_B(v) \geq \sup_u \{t(\pi_{Y|X}(v, u), \mu_A(u))\} \qquad (3.1.4)$$

The conditional possibility distribution $\pi_{Y|X}$ can be given as:

$$\pi_{Y|X}(u, v) = \begin{cases} 1 & \text{if } \mu_A(u) \leq \mu_B(v) \\ \mu_B(v) & \text{if } \mu_A(u) > \mu_B(v) \end{cases} \qquad (3.1.5)$$

if the t-norm selected is the minimum, and

$$\pi_{Y|X}(u, v) = \begin{cases} 1 & \text{if } \mu_A(u) = 0 \\ \min(1, \frac{\mu_B(v)}{\mu_A(u)}) & \text{if } \mu_A(u) > 0 \end{cases} \qquad (3.1.6)$$

if the t-norm selected is the product. For further details see Dubois and Prade (1988).

Even though all fuzzy rules used in this book are formulated in an IF - THEN form, the membership of the response will be calculated in a much simpler manner. For this purpose, the degree of fulfillment of a fuzzy rule is introduced.

3.1.3   Degree of fulfillment of a fuzzy rule

The applicability of a crisp rule is a binary (0-1) function of fulfillment of the conditions of the rule. The applicability of a fuzzy rule is a non-binary function but a certain degree in the interval [0,1] depending on the conditions of the rule. This is very similar to the definition of fuzzy sets; however, for rules, truth values of more complex statements have to be defined. Obviously the "truth grade" or "truth value" of a certain rule has to depend on the arguments to which the rule is to be applied. The truth value is not a qualitative statement on the accuracy of a rule, but it is a degree to which the rule can be applied to a particular case. In the above example, if it is $5^\circ C$ and I have to walk 500 m, the question posed is: how true is the rule of Example 3.1 (or to what degree is it applicable)?

Definition 3.1 *The truth value corresponding to the fulfillment of the conditions of a rule for given premises* $(a_1, \ldots, a_K)$ *is called the degree of fulfillment (DOF) of that rule.*

Since the DOF depends on the premise vector $(a_1, \ldots, a_K)$, the DOF of rule i is denoted $D_i(a_1, \ldots, a_K)$.

The "truth grade" of a non-fuzzy rule can be calculated by using the basic logical operator formulas given in Table 3.1. In analogy with the operators defined in Table 3.1, the DOF's of the basic logical operators have to be defined. As already pointed out in Chapter 2, in the generalization of classical set, theoretical operations are not unique. Therefore, there are several possibilities to define the DOF for basic logical operators, as developed below.

The first operator is the DOF of the complement of a fuzzy statement A_1:

$$\nu(\text{NOT } A_1) = 1 - \mu_{A_1}(a_1) \qquad (3.1.7)$$

For binary operators, there are several possibilities to define the DOF. The two most common ones are the product inference and the min-max inference.

Definition 3.2 *The DOF of the product inference for the basic logical operators is:*

$$\nu(A_1 \ AND \ A_2) = \mu_{A_1}(a_1)\mu_{A_2}(a_2) \qquad (3.1.8)$$

$$\nu(A_1 \ OR \ A_2) = \mu_{A_1}(a_1) + \mu_{A_2}(a_2) - \mu_{A_1}(a_1)\mu_{A_2}(a_2) \qquad (3.1.9)$$

$$\nu(A_1 \ XOR \ A_2) = \mu_{A_1}(a_1) + \mu_{A_2}(a_2) - 2\mu_{A_1}(a_1)\mu_{A_2}(a_2) \qquad (3.1.10)$$

Definition 3.3 *The DOF of the min-max inference for the basic logical operators is:*

$$\nu(A_1 \ AND \ A_2) = \min\left(\mu_{A_1}(a_1), \mu_{A_2}(a_2)\right) \qquad (3.1.11)$$

$$\nu(A_1 \ OR \ A_2) = \max\left(\mu_{A_1}(a_1), \mu_{A_2}(a_2)\right) \qquad (3.1.12)$$

$$\nu(A_1 \ XOR \ A_2) =$$

$$= \max\left(\min(1 - \mu_{A_1}(a_1), \mu_{A_2}(a_2)), \min(\mu_{A_1}(a_1), 1 - \mu_{A_2}(a_2))\right) \qquad (3.1.13)$$

Other pairs (t-norm, t-conorm) such as the ones defined in Table 2.1 can be used for the AND and OR operations. In our experience there appears to be no great difference in the performance of the rule systems with respect to the choice of this operator. Therefore only the product and max-min inference methods are considered in this book. Note that both the product and the min-max inference rules are generalizations of the Boolean operators: if one applies them to the simple 0-1 valued cases, both reproduce Table 3.1.

Using one of the inference methods, the DOF corresponding to the rule i can be calculated for any vector $(a_1, \ldots, a_K)$ using the inference formulas 3.1.8 to 3.1.13.

Example 3.2 *Consider the rule:*

$$If \ (1,2,3)_T \ AND \ ((1,2,6)_T \ OR \ (4,5,7)_T \ )$$

$$AND \ (2.5, 4, 4.5)_T \ then \ (0,1,2)_T$$

the DOF corresponding to the vector $(a_1, a_2, a_3, a_4) = (1.5, 4, 4.8, 3)$ *is calculated from the individual membership values, which are* $0.5, 0.5, 0.8$ *and* 0.333. *Using the product inference method the DOF is:*

$$0.5 \cdot (0.5 + 0.8 - 0.5 \cdot 0.8) \cdot 0.333 = 0.150$$

The min-max inference leads to the fulfillment grade:

$$\min (0.5, \max(0.5, 0.8), 0.333) = 0.333$$

With the vector $(a_1, a_2, a_3, a_4) = (2, 2, 5, 3)$ *the memberships are* $1, 1, 1,$ *and* 0.333. *In this case both DOF-s equal* 0.333. *In this second case three arguments have a membership value of 1. The fulfillment grade of the rule using the product inference has increased while that using the min-max inference has remained the same.*

This illustrates that the product inference rule accounts for the fulfillment of all arguments in contrast to the min-max inference where only the limiting or extreme argument is considered. Another difference between the min-max inference and the product inference is in the handling of repeated arguments. For the min-max inference one has:

$$\nu(A \text{ AND } (A \text{ OR } B)) = \nu(A)$$

$$\nu(A \text{ OR } (A \text{ AND } B)) = \nu(A)$$

This is in full conformity with Boolean logic. In contrast the product inference yields:

$$\nu(A \text{ AND } (A \text{ OR } B)) \leq \nu(A)$$

$$\nu(A \text{ OR } (A \text{ AND } B)) \geq \nu(A)$$

This means that the repeated usage of a given argument in the same rule might influence the DOF of the result. However, if rules are formulated with the single AND conjunction as suggested in the next section, the repeated use of the same argument has no reason to occur. We refer the interested reader to works by Zadeh (1972, 1983), Yager (1980) and Dubois and Prade (1980b, 1991) for further developments on this topic.

For the purposes of this book, besides the usual binary logical operators "AND" and "OR", the following multivariate operators can be defined on elements of a set, corresponding to "soft" statements like

"MOST OF THE PROPOSITIONS" are true or "AT LEAST A FEW" are true. These statements can be formulated as

$$D_i = F\left(\mu_{A_{i,1}}(a_1) \ldots \mu_{A_{i,K}}(a_K)\right) \qquad (3.1.14)$$

where the operator is defined as:

$$D_i = \gamma DOF\left(A_{i,1} \ OR \ \ldots \ OR \ A_{i,K}\right) +$$

$$+(1-\gamma)DOF\left(A_{i,1} \ AND \ \ldots \ AND \ A_{i,K}\right) \ 0 \leq \gamma \leq 1$$

$0 \leq \gamma \leq 1$. Thus using the product inference rule yields:

$$D_i = \gamma F_o\left(\mu_{A_{i,1}}(a_1), \ldots, \mu_{A_{i,K}}(a_K)\right) +$$

$$+ (1-\gamma)F_a\left(\mu_{A_{i,1}}(a_1), \ldots, \mu_{A_{i,K}}(a_K)\right) \qquad (3.1.15)$$

with F_o being the "OR" function and F_a, the "AND" function. For example taking the "OR" of the product inference yields:

$$F_o\left(x_1, x_2\right) = x_1 + x_2 - x_1 x_2, \quad (x_1, x_2)\epsilon R^2 \qquad (3.1.16)$$

For R variables, $(x_1, \ldots, x_R)$ F_o is defined recursively as:

$$F_o\left(x_1, \ldots, x_R\right) = F_o\left(F_o(x_1, \ldots, x_{R-1}), x_R\right) \qquad (3.1.17)$$

The "AND" function F_a can be defined as in the product inference:

$$F_a\left(x_1, \ldots, x_R\right) = \prod_{r=1}^{R} x_r \qquad (3.1.18)$$

Other inference methods such as maximum inference can also be used for calculating the functions F_a and F_o. The value of γ plays a balancing rule between the "AND" operator which may be too stringent ($\gamma = 0$) and the "OR" operator, which may be too lenient, that is, possess too little discriminant power. ($\gamma = 1$). A problem with the operator defined in Eq. 3.1.15 is that in case there exists some indices k and l such that $\mu_{A_{i,k}}(a_k) = 1$ and $\mu_{A_{i,l}}(a_l) = 0$ then $D_i = \gamma$ independently of all other arguments. Caution must thus be exercised when using Eq. 3.1.15.

To overcome this problem, we can be define the operator "MOST OF" using the p-norm $(p \geq 1)$:

$$D_i = 1 - \left(\sum_{k=1}^{K} \frac{1}{K} \left(1 - \mu_{A_{i,k}}(a_k) \right)^p \right)^{\frac{1}{p}} \qquad (3.1.19)$$

The corresponding "AT LEAST A FEW" is defined as:

$$D_i = \left(\sum_{k=1}^{K} \frac{1}{K} \left(\mu_{A_{i,k}}(a_k) \right)^p \right)^{\frac{1}{p}} \qquad (3.1.20)$$

The exponent $p \geq 1$ should be properly selected in either formula so as to reflect the compensatory effect between arguments $A_{i,k}$ for $k = 1, \ldots, K$. The value $p = 1$ corresponds to perfect compensation and $p = \infty$, to no compensation (Bardossy and Duckstein, 1992). In terms of membership function values, the higher is p the more the "MOST OF" operator gets close to the minimum of the values and the more the "AT LEAST A FEW", to the maximum.

The difference between the operators "AND" "OR" "MOST OF" and "AT LEAST A FEW" is illustrated by the following numerical example:

Example 3.3 *Suppose we have 5 rule arguments. For these we have* $\nu_1 = \mu_{A_1}(a_1) = \nu_2 = \mu_{A_2}(a_2) = 0.9, \nu_3 = \mu_{A_3}(a_3) = 0.5, \nu_4 = \mu_{A_4}(a_4) = 0.1$ *and* $\nu_5 = \mu_{A_5}(a_5) = 0.0.$ *Then if the arguments are coupled by the "AND" statement the degree of fulfillment of the rule is 0 as:*

$$0.9 \cdot 0.9 \cdot 0.5 \cdot 0.1 \cdot 0.0 = 0.0$$

The "OR" statement is calculated by Eqs. 3.1.16 and 3.1.17 as:

$$(0.9 + 0.9 - 0.9 \cdot 0.9) + 0.5 - (0.9 + 0.9 - 0.9 \cdot 0.9) \cdot 0.5 + 0.1-$$

$$((0.9 + 0.9 - 0.9 \cdot 0.9) + 0.5 - (0.9 + 0.9 - 0.9 \cdot 0.9) \cdot 0.5) \cdot 0.1 = 0.9955$$

If one takes "MOST OF" defined with $\gamma = 0.25$ *the DOF is calculated as:*

$$0.25 \cdot 0.9955 + 0.75 \cdot 0.0 = 0.248875$$

For "AT LEAST A FEW" one can take $\gamma = 0.75$ getting the DOF

$$0.75 \cdot 0.9955 + 0.25 \cdot 0.0 = 0.746625$$

The p-norm type combination with $p = 2$ yields for the "MOST OF" a DOF (Eq. 3.1.14)

$$1 - \left(\frac{(1-0.9)^2 + (1-0.9)^2 + (1-0.5)^2 + (1-0.1)^2 + (1-0.0)^2}{5} \right)^{\frac{1}{2}} = 0.355$$

and for "AT LEAST A FEW" 0.6132 (Eq. 3.1.15).

$$\left(\frac{0.9^2 + 0.9^2 + 0.5^2 + 0.1^2 + 0.0^2}{5} \right)^{\frac{1}{2}} = 0.613$$

Note that both definitions for the "MOST OF" and "AT LEAST A FEW" operators deliver intuitively acceptable values in contrast to the pure "AND" and "OR" operators. A direct formulation of "MOST OF" and "AT LEAST A FEW" using only the elementary operators "AND" and "OR" would be extremely difficult.

Example 3.4 *Consider a central business district with a large number of signalized intersections — say about 100. The city traffic engineer has decided to use the "green wave" type of control in off-peak hours, that is, when traffic at intersections is mostly normal, and an all-green north-south/all-red east-west (and vice versa) type of control called all or nothing, when "most intersections have long queues". Here the phrase between quotation marks has a fuzzy definition. In principle it would be possible to model this problem with a mix of AND and OR rules. However, if traffic intensity is taken at 3 levels (low, medium, high) at each intersection one might obtain up to 3^{100} rules. It is thus appropriate to control the traffic with the above "MOST OF" and "AT LEAST A FEW" type of rules. An example of a rule could be: If there is a small queue "at most of the" intersections and a long one at "at least a few", switch to all or nothing control.*

3.2   Combination of fuzzy rule responses

Fuzzy rules are usually formulated so that several rules can be applied to the same situation expressed as a vector of premises. These rules have not only different consequences but also different DOF-s given for the same input $(a_1, \ldots, a_K)$. (For example the product inference method is applied.) Therefore, the overall response that can be derived from the rule system has to be a combination of a few individual rule responses, that takes into consideration the individual DOF-s.

Assume that the technique for determining the DOF of rule i $\nu_i = D_i(a_1, \ldots, a_K)$ given any vector $(a_1, \ldots, a_K)$ has been selected. Let B_i be the response corresponding to rule i. The problem is now to define a fuzzy set B on the basis of the (B_i, ν_i).

$$B = \mathcal{C}\left((B_1, \nu_1), \ldots, (B_I, \nu_I)\right) \qquad (3.2.1)$$

where $\mathcal{C}$ denotes the so-called combination operator. It is supposed that all rules have consequences which are fuzzy subsets of the same set.

Note that this (fuzzy) response set depends on the individual fulfillment grades and can thus also be written as a function of the premise vector $(a_1, \ldots, a_K)$

$$\mathcal{C}\left((B_1, \nu_1), \ldots, (B_I, \nu_I)\right) = B(a_1, \ldots, a_K) \qquad (3.2.2)$$

There are several possibilities to combine rule responses (Dubois and Prade 1991). The most common ones are the minimum, maximum and additive combination methods as presented below.

The response of a rule depends on its DOF which can be taken into account either by multiplying each of the response membership functions by the corresponding DOF or by taking the minimum of the DOF value and the membership function value. Both approaches appear to be used in most of the combination methods.

3.2.1   Minimum combinations

The minimum combination method tries to find a combined rule response which is at least to a certain level in agreement with all applicable rules. A response that has zero membership for any of the applicable rules should also have a zero membership in the combined response. Thus, the philosophy for combining fuzzy responses considers only those elements as possible consequences that have a positive membership for all rules with a positive DOF (agreement preservation of possible responses). This technique is the so-called *minimum combination*.

Definition 3.4 *The minimum combination of responses* (B_i, ν_i) *is the fuzzy set B with the membership function:*

$$\mu_B(x) = \min_{\nu_i > 0} \nu_i \mu_{B_i}(x) \qquad\qquad (3.2.3)$$

where $\mu_{B_i}(x)$ *is the membership function of x in fuzzy set* B_i *for rule i.*

One can also crest the membership functions instead of multiplying them by the fulfillment grade. This results to the *cresting minimum combination*.

Definition 3.5 *The cresting minimum combination of responses* (B_i, ν_i) *is the fuzzy set B with the membership function:*

$$\mu_B(x) = \min_{\nu_i > 0} \min\left(\nu_i, \mu_{B_i}(x)\right) \qquad\qquad (3.2.4)$$

A disadvantage of the minimum combination methods, such as the ones defined by Eqs. 3.2.3 and 3.2.4 is that any disagreement impairs the usage of the rule system. In other words, if the rules are not carefully constructed then it may happen that the combination of the responses leads to: $B(a_1, \ldots, a_K)$ being empty $(\mu_B(x) = 0)$ for some $(a_1, \ldots, a_K)$ vectors. Therefore, use of minimum combinations requires more care than that of other combination methods.

3.2.2 Maximum combinations

To avoid the aforementioned problem of a possible disagreement, one can define a response combination method for which an outcome b in the set of possible responses, becomes possible (having a positive membership value) if there is at least one rule with positive DOF for which that outcome was a possible response. In this case only an agreement on the impossible responses is required (agreement preservation of impossible responses). In other words if an outcome has zero membership for all rule responses for which the rule has a positive DOF then the outcome also has a zero membership in the combined response. These requirements may be fulfilled by using the *maximum combination method* (Mamdani 1977).

Definition 3.6 *The maximum combination of responses* (B_i, ν_i) *is the fuzzy set B with the membership function:*

$$\mu_B(x) = \max_{i=1,\ldots,I} \nu_i \mu_{B_i}(x) \qquad\qquad (3.2.5)$$

where $\mu_{B_i}(x)$ is the membership function of the fuzzy set B_i.

One can also crest the membership functions instead of multiplying them by the fulfillment grade.

Definition 3.7 *The cresting maximum combination of responses (B_i, ν_i) is the fuzzy set B with the membership function:*

$$\mu_B(x) = \max_{i=1,\dots,I} \min\left(\nu_i, \mu_{B_i}(x)\right) \qquad (3.2.6)$$

The maximum combination methods tolerate disagreements, but they do not emphasize eventual agreements — the event of two rules giving the same result does not induce an increase of the membership function of the response, thus it has no effect on the credibility of the result. If for example the rules represent expert opinions this insensitivity to the proportion of experts agreeing is not a desirable property (Bárdossy et al., 1993a).

Furthermore the maximum combinations overemphasize rules with very vague responses. For example, a rule
"If A_1 is anything and A_2 is anything than B can be anything"
would dominate all other rules because it would exhibit both high DOF's and a high membership function $\mu_{B_i}(x)$.

3.2.3   Additive combinations

As a possible compromise between the minimum and maximum combination methods, one may select one of three possible types of additive combinations, namely the weighted sum, the normed weighted sum and the cresting versions of these two methods. Additive combinations have been proposed by Kosko (1985).

Definition 3.8 *The weighted sum combination of responses (B_i, ν_i) is the fuzzy set B with the membership function:*

$$\mu_B(x) = \frac{\sum_{i=1}^{I} \nu_i \mu_{B_i}(x)}{\max_u \sum_{i=1}^{I} \nu_i \mu_{B_i}(u)} \qquad (3.2.7)$$

The division by the maximum of the sum is required to ensure that the resulting membership function is not greater than 1.

In general one can state that a rule is better if its consequence is more specific. A rule with the response "anything" has no value at all. To take this specificity into account in the case when consequences are very different in vagueness, another combination method can be defined, namely, the *normed weighted sum combination*.

Definition 3.9 *The normed weighted sum combination of responses* (B_i, ν_i) *is the fuzzy set B with the membership function:*

$$\mu_B(x) = \frac{\sum_{i=1}^{I} \nu_i \beta_i \mu_{B_i}(x)}{\max_u \sum_{i=1}^{I} \nu_i \beta_i \mu_{B_i}(u)} \tag{3.2.8}$$

where

$$\frac{1}{\beta_i} = \int_{-\infty}^{+\infty} \mu_{B_i}(x)\, dx \tag{3.2.9}$$

or else, in the case of a discrete response set $B_i = \{(b_1, \mu_i(1)), \ldots, (b_J, \mu_i(J))\}$

$$\frac{1}{\beta_i} = car(B_i) = \sum \mu_i(j) \tag{3.2.10}$$

In this case each consequence B_i is assigned a unit weight but rules have different weights. Eqs. 3.2.8, 3.2.9 and 3.2.10 mean that rules with crisper answers carry greater weight than rules with very fuzzy (uncertain) answers.

The cresting version of the above additive combination methods can also be defined:

Definition 3.10 *The cresting weighted sum combination of responses* (B_i, ν_i) *is the fuzzy set B with the membership function:*

$$\mu_B(x) = \frac{\sum_{i=1}^{I} \min(\nu_i, \mu_{B_i}(x))}{\max_u \sum_{i=1}^{I} \min(\nu_i, \mu_{B_i}(x))} \tag{3.2.11}$$

Here again the division by the maximum of the sum is required to ensure that the resulting membership function is not greater than 1.

The normed version of this combination method can be defined in an analogous manner as:

Definition 3.11 *The cresting normed weighted sum combination of responses* (B_i, ν_i) *is the fuzzy set B with the membership function:*

$$\mu_B(x) = \frac{\sum_{i=1}^{I} \beta_i \min(\nu_i, \mu_{B_i}(x))}{\max_u \sum_{i=1}^{I} \beta_i \min(\nu_i, \mu_{B_i}(x))} \qquad (3.2.12)$$

The β_i values are defined in Eqs. 3.2.9 and 3.2.10. The right hand side of Eq. 3.2.12 is divided by the maximum of the numerator to ensure that the resulting membership function is not greater than 1.

Additive combinations of rules take the agreement of responses into account since by virtue of adding the membership functions, the membership of elements with such agreements increases. On the other hand no responses that were impossible for each rule with positive DOF may become possible by additive combination methods. The support of the result of the additive and the maximum combination method is the same.

3.2.4 Properties of the combination methods

The combination methods should fulfill a number of rational requirements. The most important ones are listed here:

1. Idempotency: If a response is combined with itself then the combined response should not be altered:

$$C((B_1, \nu_1), (B_1, \nu_1)) = C((B_1, \nu_1))$$

    Note that this property does not mean that the response should be equal to the unique response. Combination methods may alter single responses, for example, by cresting.

2. Associativity :

$$C((B_1, \nu_1), C((B_2, \nu_2), (B_3, \nu_3))) = C(C((B_1, \nu_1), (B_2, \nu_2)), (B_3, \nu_3))$$

3. Symmetry :

$$C((B_1, \nu_1), (B_2, \nu_2)) = C((B_2, \nu_2), (B_1, \nu_1))$$

The associativity and the symmetry ensure that the combination of the rule responses can be calculated in any order without altering the final result. All the above mentioned combination methods do fulfill these conditions. A discussion of further combination methods and their other properties can be found in Bárdossy et al. (1993a).

Note that even if all B_i-s are fuzzy numbers, the fuzzy set B obtained as their combination is not necessarily a fuzzy number. Either or both the normality and the convexity assumptions may be violated. However the minimum combination method ensures the convexity of the response if all B_i-s are convex, and the additive combinations yield by definition normal fuzzy responses.

3.2.5   Other combination methods

There are a great number of other possibilities to combine fuzzy responses. Even though in particular cases these methods deliver results that are more reasonable than the previously mentioned ones, in most of the cases the differences are not significant. Therefore, we restrict ourselves to the above combination methods.

3.2.6   Comparison of the combination methods

The difference between the combination methods and their advantages and disadvantages are pointed out and illustrated in this section.

The difference between the three combination techniques minimum, maximum and additive is that in the first two cases no allowance is made to reflect how often a consequence is assigned as being possible while in the case of the additive combinations this frequency plays an important role.

A further advantage of the additive combination methods (Defs. 3.8 and 3.9) is that all rules can be formulated with the AND operator only, without causing any change in the consequence. This property can be derived from the following two propositions, which are only stated. The proofs are found in the Appendix.

Proposition 3.1 *If the product inference rules (Eq. 3.1.8-3.1.10) and an additive combination method (Def. 3.8 or 3.9) are used then the rule*

$$\text{If } A_1 \text{ OR } A_2 \text{ then } B$$

can be replaced by three rules:

$$\text{If } A_1 \text{ AND } A_2 \text{ then } B$$

$$\text{If } A_1 \text{ AND (NOT } A_2) \text{ then } B$$

$$\text{If (NOT } A_1) \text{ AND } A_2 \text{ then } B$$

without changing the consequence.

Proposition 3.2 *If the product inference rules (Eqs. 3.1.8-3.1.10) and an additive combination method (Def. 3.8 or 3.9) are used then the rule*

$$\text{If } A_1 \text{ XOR } A_2 \text{ then } B$$

can be replaced by two rules:

$$\text{If } A_1 \text{ AND (NOT } A_2) \text{ then } B$$

$$\text{If (NOTA}_1) \text{ AND } A_2 \text{ then } B$$

without changing the consequence.

The same is not true in the case of other inference or combination methods, as shown in the following (counter) example.

Example 3.5 *Using the min-max inference method (Eq. 3.1.11-3.1.13) and the maximum combination method (Def. 3.6) with $\mu_{A_1}(a_1) = 0.7$ and $\mu_{A_2}(a_2) = 0.6$ the rule*

$$\text{If } A_1 \text{ OR } A_2 \text{ then } B_1$$

has a DOF=0.7. Thus, the response is $0.7\mu_{B_1}(x)$. The three rules:

$$\text{If } A_1 \text{ AND } A_2 \text{ then } B_1$$

$$\text{If } A_1 \text{AND (NOT } A_2) \text{ then } B_1$$

$$\text{If (NOTA}_1) \text{ AND } A_2 \text{ then } B_1$$

have the respective DOF-s 0.6, 0.4 and 0.3. The maximum combination is thus $0.6\mu_{B_1}(x)$, which is different from $0.7\mu_{B_1}(x)$.

Similar counterexamples can be found using combination methods other than the additive ones in conjunction with any inference method.

In particular, Propositions 1 and 2 show that in the case of the product inference used in conjunction with the additive combination method each rule system can be formulated by using the AND operator only. This simplifies the construction of the rule system — and the evaluation of the DOFs can be done in any order. Further advantages of this combination method will be demonstrated in the next section in conjunction with defuzzification techniques.

Example 3.6 *Consider the following rule responses:*
Case 1: Rule 1 has the DOF $\nu_1 = 0.4$ and the consequence $B_1 = (0, 2, 4)_T$
Rule 2 has the DOF $\nu_2 = 0.5$ and the consequence $B_2 = (3, 4, 5)_T$ No other rules apply ($\nu_i = 0$ if $i > 2$).

As the supports of B_1 and B_2 have the intersection $[3, 4]$ the minimum is 0 outside this interval. The minimum combination is thus obtained by taking the minimum of the two membership functions:

$$\min\left(0.4\frac{4-x}{2}, 0.5(x-3)\right)$$

on $[3, 4]$. As the equation

$$0.4\frac{4-x}{2} = 0.5(x-3)$$

has the solution $\frac{23}{7}$, the minimum of the two functions is:

$$\mu_B(x) = \begin{cases} 0.5(x-3) & \text{if } 3 < x \leq \frac{23}{7} \\ 0.4\frac{4-x}{2} & \text{if } \frac{23}{7} < x < 4 \end{cases}$$

The cresting minimum combination is obtained by taking the minimum of the two membership functions μ_{B_1} μ_{B_2} and the two fulfillment grades ν_1 and ν_2. The equation

$$\frac{4-x}{2} = x - 3$$

has the solution $\frac{10}{3}$, the minimum of the two functions and the two constants are thus found to be:

$$\mu_B(x) = \begin{cases} x-3 & \text{if } 3 < x \leq \frac{10}{3} \\ \frac{4-x}{2} & \text{if } \frac{10}{3} < x < 4 \end{cases}$$

Figure 3.2 shows the membership functions of the consequences of the minimum and cresting minimum combination methods.

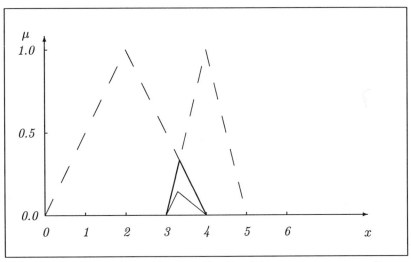

Figure 3.2. **Membership functions of the consequence** B **using combination methods** *minimum combination* **(thin lines) and** *cresting minimum combination* **(thick lines), the individual rule consequences are dashed.**

The maximum combination is obtained by taking the maximum of the two membership functions, which yields the membership function:

$$\mu_B(x) = \begin{cases} 0.4\frac{x}{2} & \text{if } 0 < x \le 2 \\ 0.4\frac{4-x}{2} & \text{if } 2 < x \le \frac{23}{7} \\ 0.5(x-3) & \text{if } \frac{23}{7} < x \le 4 \\ 0.5(5-x) & \text{if } 4 < x \le 5 \end{cases}$$

The cresting maximum combination requires more algebraic effort, as first the minimum of the fulfillment grades and then the membership functions have to be calculated.

$$\mu_B(x) = \begin{cases} \frac{x}{2} & \text{if } 0 < x \le 0.8 \\ 0.4 & \text{if } 0.8 < x \le 3.2 \\ \frac{4-x}{2} & \text{if } 3.2 < x \le \frac{10}{3} \\ (x-3) & \text{if } \frac{10}{3} < x \le 3.5 \\ 0.5 & \text{if } 3.5 < x \le 4.5 \\ (5-x) & \text{if } 4.5 < x \le 5 \end{cases}$$

Figure 3.3 shows the membership functions of the consequences of the maximum and cresting maximum combination methods.

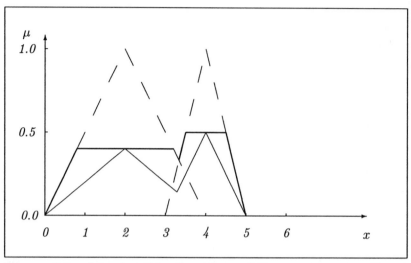

Figure 3.3. **Membership functions of the consequence B using combination methods** *maximum combination* **(thin lines) and** *cresting maximum combination* **(thick lines), the individual rule consequences are dashed.**

The weighted sum combination is obtained by first taking the sum of the two membership functions, which is

$$f(x) = \begin{cases} 0.4\frac{x}{2} & \text{if } 0 < x \le 2 \\ 0.4\frac{4-x}{2} & \text{if } 2 < x \le 3 \\ 0.4\frac{4-x}{2} + 0.5(x-3) & \text{if } 3 < x \le 4 \\ 0.5(5-x) & \text{if } 4 < x \le 5 \end{cases}$$

As the maximum of $f(x)$ is 0.5 the membership function of B is obtained as

$$\mu_B(x) = 2f(x)$$

The area under B_1 $\left(\frac{1}{\beta_1}\right)$ is 2 and that under B_2 $\left(\frac{1}{\beta_2}\right)$ is 1. Therefore the normed weighted sum combination is obtained by taking the weighted *sum function $f(x)$ first:*

$$f(x) = \begin{cases} 0.2\frac{x}{2} & \textit{if } 0 < x \leq 2 \\ 0.2\frac{4-x}{2} & \textit{if } 2 < x \leq 3 \\ 0.2\frac{4-x}{2} + 0.5(x-3) & \textit{if } 3 < x \leq 4 \\ 0.5(5-x) & \textit{if } 4 < x \leq 5 \end{cases}$$

As the maximum of $f(x)$ is 0.5 the membership function of B is obtained as

$$\mu_B(x) = 2f(x)$$

Figure 3.4 shows the membership functions of the consequences of the additive and weighted additive combination methods.

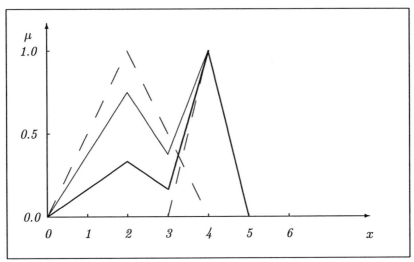

Figure 3.4. **Membership functions of the consequence B using combination methods** *weighted sum combination* **(thin lines) and** *normed weighted sum combination* **(thick lines).**

The cresting weighted sum combination is obtained by first taking the minimum of the DOF-s grade and the membership functions, and then summing the two functions:

$$f(x) = \begin{cases} \frac{x}{2} & \textit{if } 0 < x \le 0.8 \\ 0.4 & \textit{if } 0.8 < x \le 3 \\ x - 2.6 & \textit{if } 3 < x \le 3.2 \\ \frac{4-x}{2} + x - 3 & \textit{if } 3.2 < x \le 3.5 \\ \frac{4-x}{2} + 0.5 & \textit{if } 3.5 < x \le 4 \\ 0.5 & \textit{if } 4 < x \le 4.5 \\ 5 - x & \textit{if } 4.5 < x \le 5 \end{cases}$$

As the maximum of $f(x)$ is 0.75 the membership function of B is obtained as

$$\mu_B(x) = \frac{f(x)}{0.75}$$

The area under B_1 ($\frac{1}{\beta_1}$) is 2 and that under B_2 ($\frac{1}{\beta_2}$) is 1. Therefore the cresting normed weighted sum combination is obtained by taking the weighted sum function $f(x)$ first:

$$f(x) = \begin{cases} 0.5\frac{x}{2} & \textit{if } 0 < x \le 0.8 \\ 0.2 & \textit{if } 0.8 < x \le 3 \\ x - 2.8 & \textit{if } 3 < x \le 3.2 \\ 0.5\frac{4-x}{2} + x - 3 & \textit{if } 3.2 < x \le 3.5 \\ 0.5\frac{4-x}{2} + 0.5 & \textit{if } 3.5 < x \le 4 \\ 0.5 & \textit{if } 4 < x \le 4.5 \\ 5 - x & \textit{if } 4.5 < x \le 5 \end{cases}$$

As the maximum of $f(x)$ is 0.7, the membership function of B is obtained as

$$\mu_B(x) = \frac{f(x)}{0.7}$$

Figure 3.5 shows the membership functions of the consequences of the additive and weighted additive combination methods.

Case 2: Rule 1 has the DOF $\nu_1 = 0.4$ and the consequence $B_1 = (0, 2, 4)_T$ Rules 2-I have the DOF $\nu_i = 0.5$ and all have the same consequence $B_i = (3, 4, 5)_T$. In this case the minimum and maximum combinations yield exactly the same result as before. Taking $I = 6$ the weighted sum combination is obtained by taking first the sum of the membership

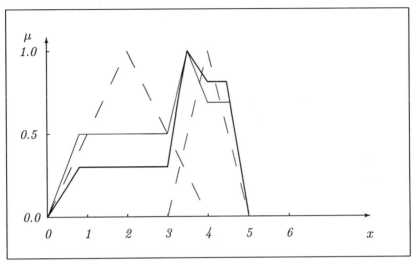

Figure 3.5. Membership functions of the consequence B using combination methods *cresting weighted sum combination* (**thin lines**) and *cresting normed weighted sum combination* (**thick lines**), the individual rule consequences are dashed.

functions, which is

$$f(x) = \begin{cases} 0.4\frac{x}{2} & \text{if } 0 < x \leq 2 \\ 0.4\frac{4-x}{2} & \text{if } 2 < x \leq 3 \\ 0.4\frac{4-x}{2} + 5 \cdot 0.5(x-3) & \text{if } 3 < x \leq 4 \\ 5 \cdot 0.5(5-x) & \text{if } 4 < x \leq 5 \end{cases}$$

As the maximum of $f(x)$ is 2.5 the membership function of B is obtained as

$$\mu_B(x) = 0.4 f(x)$$

The area under B_1 $(\frac{1}{\beta_1})$ is 2 and under B_2 $(\frac{1}{\beta_2})$ is 1. Therefore the normed weighted sum combination is obtained by taking the weighted sum function $f(x)$ first:

$$f(x) = \begin{cases} 0.2\frac{x}{2} & \text{if } 0 < x \leq 2 \\ 0.2\frac{4-x}{2} & \text{if } 2 < x \leq 3 \\ 0.2\frac{4-x}{2} + 5 \cdot 0.5(x-3) & \text{if } 3 < x \leq 4 \\ 5 \cdot 0.5(5-x) & \text{if } 4 < x \leq 5 \end{cases}$$

As the maximum of $f(x)$ is 2.5 the membership function of B is obtained as

$$\mu_B(x) = 0.4f(x)$$

Figure 3.6 shows the membership functions of the consequences of the weighted sum and normed weighted sum combination methods.

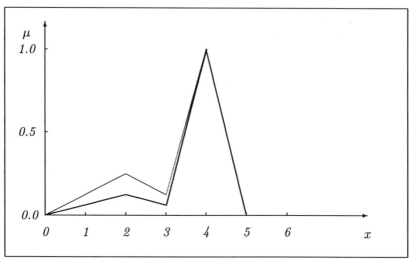

Figure 3.6. **Membership functions of the consequence B using combination methods** *weighted sum combination* **(thin lines) and** *normed weighted sum combination* **(thick lines) if $I = 6$ in Case 2.**

The cresting version of the additive combination is also different from the previous case. The cresting weighted sum yields:

$$f(x) = \begin{cases} \frac{x}{2} & \text{if } 0 < x \le 0.8 \\ 0.4 & \text{if } 0.8 < x \le 3 \\ 5(x-3) + 0.4 & \text{if } 3 < x \le 3.2 \\ \frac{4-x}{2} + 5(x-3) & \text{if } 3.2 < x \le 3.5 \\ \frac{4-x}{2} + 2.5 & \text{if } 3.5 < x \le 4 \\ 2.5 & \text{if } 4 < x \le 4.5 \\ 5(5-x) & \text{if } 4.5 < x \le 5 \end{cases}$$

As the maximum of $f(x)$ is 2.75 the membership function of B is obtained as

$$\mu_B(x) = \frac{f(x)}{2.75}$$

The area under B_1 $\left(\frac{1}{\beta_1}\right)$ is 2 and that under B_i $\left(\frac{1}{\beta_i}, i > 1\right)$ is 1. Therefore, the cresting normed weighted sum combination is obtained by taking the weighted sum function $f(x)$ first:

$$f(x) = \begin{cases} 0.5\frac{x}{2} & \text{if } 0 < x \leq 0.8 \\ 0.2 & \text{if } 0.8 < x \leq 3 \\ 5(x-3) + 0.2 & \text{if } 3 < x \leq 3.2 \\ 0.5\frac{4-x}{2} + 5(x-3) & \text{if } 3.2 < x \leq 3.5 \\ 0.5\frac{4-x}{2} + 2.5 & \text{if } 3.5 < x \leq 4 \\ 2.5 & \text{if } 4 < x \leq 4.5 \\ 5(5-x) & \text{if } 4.5 < x \leq 5 \end{cases}$$

As the maximum of $f(x)$ is 2.7 the membership function of B is obtained as

$$\mu_B(x) = \frac{f(x)}{2.7}$$

Figure 3.7 shows the membership functions of the consequences of the additive and weighted additive combination methods.

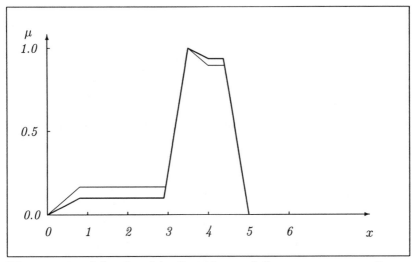

Figure 3.7. **Membership functions of the consequence B using combination methods** *cresting weighted sum combination* **(thin lines) and** *cresting normed weighted sum combination* **(thick lines) if** $I = 6$ **in Case 2.**

Note that both methods result in fuzzy numbers defined over the same support [0,5]. Furthermore, Case 1 and Case 2 yield the same fuzzy consequence B whether one uses minimum, crested minimum, maximum (Def. 3.6) or the cresting maximum combination method (Def. 3.7). This means that the much more frequent consequence B_2 in Case 2 does not have any effect on the overall consequence. This is not the case for the additive (Def. 3.8) and weighted additive (Def. 3.9) combination methods which gradually emphasize the more frequent consequence. This property is advantageous if there is no major redundancy in the rule system.

The difference between the weighted sum combination and the normed weighted sum combination lies in the treatment of the fuzziness of the consequences. The weighted sum combination does not account for the different uncertainties inherent to the consequence elements. In contrast the normed weighted sum combination places more weight on the crisper consequence.

Example 3.7 *Consider the following rule system:*
Rule 1 has the DOF $\nu_1 = 0.5$ and the consequence $B_1 = (0, 3, 6)_T$
Rule 2 has the DOF $\nu_2 = 0.5$ and the consequence $B_2 = (3, 4, 5)_T$
no other rule applies.
    As for each x:

$$\nu_1 \mu_{B_1}(x) \geq \nu_2 \mu_{B_2}(x) \tag{3.2.13}$$

the minimum combination of the two responses is

$$\mu_B(x) = \nu_2 \mu_{B_2}(x)$$

For the minima of DOF and membership function one finds:

$$\min(\nu_1, \mu_{B_1}(x)) \geq \min(\nu_2, \mu_{B_2}(x)) \tag{3.2.14}$$

Thus the cresting minimum combination is

$$\mu_B(x) = \begin{cases} 0.5(x-3) & \text{if } 3 < x \leq 3.5 \\ 0.5 & \text{if } 3.5 < x \leq 4.5 \\ 0.5(5-x) & \text{if } 4.5 < x \leq 5 \end{cases}$$

Both minimum combination methods neglect B_1 — the result is the same as it would be in the case when the first rule would have a DOF equal to zero.

    The maximum combination of the two responses is, using Eq. (3.2.13)

$$\mu_B(x) = \nu_1 \mu_{B_1}(x)$$

The cresting maximum combination is, as a consequence of Eq. (3.2.14)

$$\mu_B(x) = \begin{cases} 0.5\frac{x}{3} & \text{if } 0 < x \leq 1.5 \\ 0.5 & \text{if } 1.5 < x \leq 4.5 \\ 0.5\frac{6-x}{3} & \text{if } 4.5 < x \leq 6 \end{cases}$$

In contrast to the minimum combination methods both maximum combination methods neglect B_2 - the result is the same as it would be in the case of the second rule having a DOF equal to zero.

The weighted combination method provides a response which depends on both B_1 and B_2. The membership function of B in this case is obtained from the weighted sum of the membership functions $f(x)$

$$f(x) = \begin{cases} 0.5\frac{x}{3} & \text{if } 0 < x \leq 3 \\ 0.5\frac{6-x}{3} + 0.5(x-3) & \text{if } 3 < x \leq 4 \\ 0.5\frac{6-x}{3} + 0.5(5-x) & \text{if } 4 < x \leq 5 \\ 0.5\frac{6-x}{3} & \text{if } 5 < x \leq 6 \end{cases}$$

The maximum of $f(x)$ is $\frac{5}{6}$, thus

$$\mu_B(x) = \frac{6}{5}f(x)$$

The normed weighted sum combination also takes the different uncertainties in the responses into account. The area under B_1 is 3, that under B_2 is 1 and the weights β_i are selected as the inverses of these numbers (Eq. 3.2.9). The membership function of B in this case is obtained from the weighted sum of the membership functions $f(x)$

$$f(x) = \begin{cases} 0.5\frac{1}{3}\frac{x}{3} & \text{if } 0 < x \leq 3 \\ 0.5\frac{1}{3}\frac{6-x}{3} + 0.5(x-3) & \text{if } 3 < x \leq 4 \\ 0.5\frac{1}{3}\frac{6-x}{3} + 0.5(5-x) & \text{if } 4 < x \leq 5 \\ 0.5\frac{1}{3}\frac{6-x}{3} & \text{if } 5 < x \leq 6 \end{cases}$$

The maximum of $f(x)$ is $\frac{11}{18}$ thus

$$\mu_B(x) = \frac{18}{11}f(x)$$

Figure 3.8 shows the membership functions of the two different consequences B. Note that the membership function of the normed weighted sum is closer to the crisper response B_2. The weighted sum combination

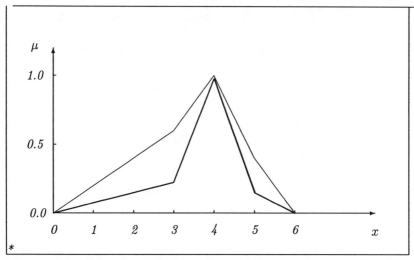

Figure 3.8. **Membership functions of the consequence** B **using combination methods** *weighted sum combination* **(thin lines) and** *normed weighted sum combination* **(thick lines)**

leads to a response in which B_1 *plays a major role. In contrast to the minimum and maximum combination methods these combinations are not "dictatorial" they do not neglect any of the responses.*

3.3   Defuzzification

A rule system applied to a certain case described by premise vector $(a_1, \ldots, a_K)$ leads to a consequence B which is usually a fuzzy set, obtained as a combination of the individual rule responses. It is often necessary to replace the fuzzy consequence B with a single crisp consequence b, for example a point prediction may be required for forecasting, decision making or control.

Definition 3.12 *The procedure that transforms a fuzzy consequence* B *into a crisp consequence* b *is called defuzzification. The defuzzification will be denoted as:*

$$b = \mathcal{D}_f(B)$$

As the consequence $B(a_1, \ldots, a_K)$ depends on the vector of premises $(a_1, \ldots, a_K)$ so does the defuzzified value; let this defuzzified consequence or rule response (rule response function) be denoted as $R(a_1, \ldots, a_K)$. By definition:

$$R(a_1, \ldots, a_K) = \mathcal{D}_f(B(a_1, \ldots, a_K))$$

There are two major cases of defuzzification depending on the universe on which the responses are defined.

1. The response is in an arbitrary or discrete numerical universe, for example categories.

2. The response is a continuous numerical variable, such as a hydrologic forecast.

These cases are treated separately in the next section.

3.3.1   Defuzzification methods

In the first case of a discrete response, a single element has to be selected. The most frequently used technique consists of selecting the element with maximal membership value of the consequence as defuzzified response. Obviously this method can also be applied in the continuous case.

Definition 3.13 *The defuzzification by maximum of the fuzzy consequence B selects the element with the maximum membership value as the representative element of the consequence set B.*

$$\mu_B(b) = \max_x \mu_B(x) \qquad (3.3.1)$$

The defuzzification by maximum is not necessarily unique, since cresting combination methods may often lead to an infinite number of elements having the maximal value as membership as illustrated for example in Fig. 3.3. Therefore this defuzzification method is not recommended in the case of continuous numerical responses.

In the continuous case if the rules provide consequences that are fuzzy numbers or fuzzy sets defined on the real line with continuous membership functions the defuzzification can be done by using a location parameter, for example, the fuzzy mean of B.

Definition 3.14 *The mean defuzzification of the fuzzy consequence B is its fuzzy mean Eq. 2.2.6:*

$$b = \mathcal{D}_f(B) = M(B) \qquad (3.3.2)$$

Another possibility is to use the fuzzy median:

Definition 3.15 *The median defuzzification of the fuzzy consequence B is its fuzzy median as given by Eq. 2.2.12:*

$$b = \mathcal{D}_f(B) = m(B) \qquad (3.3.3)$$

Its modified version as given by Eq. 2.2.15 provides an alternative to the maximum defuzzification method for discrete ordered sets.

The minimum and maximum combination methods defined by Eqs. 3.2.5 and 3.2.6 result in B-s constructed with the help of the max-min operators that make the calculation of $M(B)$ rather difficult. In the case of piecewise linear membership functions, or in the case when all individual responses are triangular or trapezoidal, the fuzzy mean of B can be calculated directly using Eq. 2.2.9. However, in order to find the breakpoints of the membership function of the combined fuzzy set B an additional algorithm has to be applied.

An advantage of the additive combination methods (Defs. 3.8 and 3.9) is that the fuzzy mean of the resulting consequence can be calculated directly. Thus, for the weighted sum combination method (Eq. 3.2.7) the fuzzy mean is:

$$
\begin{aligned}
M(B) &= \frac{\int t \sum_{i=1}^{I} \nu_i \mu_{B_i}(t)\, dt}{\int \sum_{i=1}^{I} \nu_i \mu_{B_i}(t)\, dt} = \\
&= \frac{\sum_{i=1}^{I} \nu_i \int t \mu_{B_i}(t)\, dt}{\sum_{i=1}^{I} \nu_i \int \mu_{B_i}(t)\, dt} = \\
&= \frac{\sum_{i=1}^{I} \nu_i \frac{1}{\beta_i} M(B_i)}{\sum_{i=1}^{I} \nu_i \frac{1}{\beta_i}}
\end{aligned}
\qquad (3.3.4)
$$

Here β_i is the inverse of the area under the membership function (or in the discrete case the inverse of the cardinality) as defined in Eq. 3.2.9. Similarly, using the normed weighted sum combination method (Eq. 3.2.8) the fuzzy mean is:

$$
\begin{aligned}
M(B) &= \frac{\int t \sum_{i=1}^{I} \nu_i \beta_i \mu_{B_i}(t)\, dt}{\int \sum_{i=1}^{I} \nu_i \beta_i \mu_{B_i}(t)\, dt} = \\
&= \frac{\sum_{i=1}^{I} \nu_i \beta_i \int t \mu_{B_i}(t)\, dt}{\sum_{i=1}^{I} \nu_i \beta_i \int \mu_{B_i}(t)\, dt} =
\end{aligned}
$$

$$= \frac{\sum_{i=1}^{I} \nu_i M(B_i)}{\sum_{i=1}^{I} \nu_i} \qquad (3.3.5)$$

This means that in the case of weighted sum combination and mean defuzzification, rules can be formulated simply as:

$$\text{If } A_{i,1} \text{ AND } A_{i,2} \text{ AND} \ldots \text{AND } A_{i,K} \text{ then } (M(B_i), \beta_i) \qquad (3.3.6)$$

There is no need to define the entire rule response B_i as a fuzzy set. Its fuzzy mean $M(B_i)$ and the area under the membership function β_i are sufficient for the calculation of the rule response. The β_i value acts as a further weight — the higher its value the greater the influence of the rule on the final composite defuzzified response. Note that 3.3.6 corresponds to the method of fuzzy control of Sugeno (1985) and Sugeno and Nishida (1985).

The normed weighted sum combination and the mean defuzzification lead to an even simpler rule formulation:

$$\text{If } A_{i,1} \text{ AND } A_{i,2} \text{ AND} \ldots \text{AND } A_{i,K} \text{ then } M(B_i) \qquad (3.3.7)$$

Here only the fuzzy mean of the rule response has to be specified. This property and the fact that all rules can be formulated using only with the logical AND and the simple combination of the rule responses makes this pair combination method-defuzzification especially attractive for practical applications. These properties are advantageous in rule assessment, too. However, for some applications, the other methods might be more relevant.

The cresting additive combinations (Definitions 3.10 and 3.11) require somewhat greater computational efforts for calculating the defuzzified response. In the case when the responses are triangular fuzzy numbers, the calculations can be simplified substantially.

Example 3.8 *For the fuzzy consequences defined in Example 3.6 the results of defuzzification by the methods described above have been calculated. The different responses thus obtained are displayed in Table 3.2*

It may be noted that the result of the cresting maximum combination followed by the maximum defuzzification does not lead to a unique answer. A wide range of defuzzified responses, from 2.638 to 4.5 is found in both cases. The minimum combination yields defuzzified values very close to each other for all defuzzification methods.

Table 3.2.   The defuzzified responses for Examples 3.6 and 3.7

Combination method	Case 1		
	Defuzzification method		
	Maximum	Mean	Median
Minimum	3.287	3.429	3.402
Cresting minimum	3.333	3.444	3.423
Maximum	4.000	2.731	2.638
Cresting maximum	[3.5, 4.5]	2.676	2.729
Weighted sum	4.000	2.769	2.775
Normed weighted sum	4.000	3.111	3.309
Cresting w.s.	3.500	2.739	2.743
Cresting n.w.s	3.500	3.089	3.279
Combination method	Case 2		
	Defuzzification method		
	Maximum	Mean	Median
Minimum	3.287	3.429	3.402
Cresting minimum	3.333	3.444	3.423
Maximum	4.000	2.731	2.638
Cresting maximum	[3.5, 4.5]	2.676	2.729
Weighted sum	4.000	3.515	3.562
Normed weighted sum	4.000	3.724	3.637
Cresting w.s.	3.500	3.491	3.537
Cresting n.w.s	3.500	3.707	3.618

There is almost no difference between the additive and the cresting additive responses followed by the mean or median defuzzification.

A defuzzification using distance measures by taking the closest crisp number to a fuzzy set could also be considered. The problem with this is that the distances are only defined for fuzzy numbers and can only be extended to act on convex fuzzy sets on the real line. Therefore, to use a distance-based defuzzification, the combination methods should provide convex responses. This can be obtained by taking the convex hull of the result of any combination method (the "minimal" convex fuzzy set containing the combined fuzzy response). However this procedure requires additional effort, and the responses are not that different — thus it is not really worth the effort to develop such a scheme in detail.

A few further examples will be shown to illustrate possible differences between the defuzzification methods. The different inference methods, combination techniques and defuzzification procedure provide a great number of possibilities to evaluate fuzzy rules. The natural question arises about which formulation should be used for a selected application. The product inference with additive combinations and fuzzy mean defuzzification offer the simplest rule formulation (using AND only) and rule evaluation; therefore, this may be the preferred approach. In the next chapter, further results that support this choice are reported.

3.4   Case of fuzzy premises

Until now it was assumed that the rules are applied to exactly known crisp premises. However often there are cases where in addition to the rules arguments and responses being fuzzy, the premises of the system are also uncertain and can be described using fuzzy sets. This is very typical in human decision problems where both the inputs and the rules are inexactly known — but still actions have to be taken. Whereas in the usual case only the truth of a statement "the element a is A" has to be assessed, here the truth of the statement the fuzzy set "$\tilde{a}$ is A" also has to be determined.

Example 3.9 *Consider the problem of the number of people on a Sunday at a particular beach. It is clear that the daily maximum temperature has a great influence on that figure. One could assess rules to describe this relationship using past observations. However if one wants to use these rules for the coming weekend, the uncertainty of the weather forecast also has to be taken into account. Thus the rules input consists of* *a fuzzy forecast instead of the exactly measured temperatures that have been used to establish the rules.*

As mentioned before, fuzzy rule systems are formulated in an ambiguous way, so that usually more than one rule can be applied to a certain case. The use of fuzzy premises increases this ambiguity.

There are (at least) two ways to cope with fuzzy inputs. The first one is to handle them similarly to crisp premises — and get a consequence B which is defuzzified as before: we call this the fuzzy premises — crisp response technique. The second possibility is to apply the rule system to the different level sets of the fuzzy premises yielding finally a two dimensional fuzzy set B — which, after partial defuzzification as shown below, results in a one dimensional fuzzy output: this is the so-called fuzzy premises — fuzzy response technique.

3.4.1 Fuzzy premises — crisp response

Let $\mu_A(x)$ be the membership function of A and $\mu_{\tilde{a}}(x)$, that of the fuzzy premise $\tilde{a}$. Using the extension principle to the function $DOF(a)$, the DOF of a statement is

$$\nu(\tilde{a} \text{ is } A) = \max_x \left(\min(\mu_{\tilde{a}}(x), \mu_A(x)) \right) \qquad (3.4.1)$$

If $\tilde{a}$ and A are both fuzzy numbers, the evaluation of Eq. 3.4.1 can be simplified here, as:

$$\nu(\tilde{a} \text{ is } A) = \max_x \left(\mu_{\tilde{a}}(x) \text{ such that } \mu_{\tilde{a}}(x) = \mu_A(x) \right) \qquad (3.4.2)$$

Example 3.10 *The fuzzy set of tall men can be defined as a triangular fuzzy number $A = (175, 185, 210)_T$ cm. A person with an estimated height of $\tilde{a} = (174, 179, 184)_T$ cm has to be assigned a DOF of the statement "$\tilde{a}$ is A" meaning "this person is tall". From the definition in Eq. 3.4.2 this value is found to be 0.625.*

Using Eqs. 3.4.1 and 3.4.2 fuzzy premises can be incorporated into rule systems. It is usually not possible to define a crisp premise that is equivalent to a fuzzy premise in the sense of resulting in the same DOF for all individual rules. In the case of fuzzy premises there are usually more rules with positive DOFs than for crisp premises.

3.4.2 Fuzzy premises — fuzzy responses

The second possibility to deal with fuzzy premises is to apply the rule system at each point of a selected level set of $\tilde{a}$. Specifically, for each $(a_1, \ldots, a_K)$, a fuzzy response set $B(a_1, \ldots, a_K)$ with the membership vector $(\mu_1(a_1), \ldots, \mu_K(a_K))$ is defined, where $\mu_k(a_k)$ is the membership function of $\tilde{a}_k$. Here again one is faced with the problem of defuzzifying a multidimensional fuzzy set, which can be done by performing a defuzzification in each dimension. Using the rule for the membership function of a Cartesian set (Def. 2.16) yields:

$$\mu(R(a_1, \ldots, a_K)) = \min_k(\mu_1(a_1), \ldots, \mu_K(a_K)) \qquad (3.4.3)$$

This operation results in a one dimensional fuzzy set, which can then be regarded as the fuzzy consequence of the fuzzy premise.

A quite usual occurrence is that data are incomplete; for example, certain measurements may be missing. Thus, in a case of medical diagnosis not all possible observations have been taken, but on the basis of incomplete data, responses still have to be found. Taking the missing a_k-s as fuzzy sets with the rectangular membership function

$$\mu_k(a_k) = 1 \quad \text{if} \quad a_{k\,\min} \leq a_k \leq a_{k\,\max} \qquad (3.4.4)$$

the rule system with fuzzy premises can be applied. Using the first method a crisp consequence can be obtained, the second one delivers a fuzzy consequence reflecting the uncertainty resulting from the missing argument(s). The price to be paid for this additional information is that much more difficult calculations are needed to find the result. The following example illustrates the two approaches.

Example 3.11 *Consider the two rules*

$$\text{If } (1, 2, 3)_T \text{ AND } (2, 3, 6)_T \text{ then } (0, 1, 2)_T$$

having the fuzzy mean 1 as the consequence, and

$$\text{If } (2, 3, 5)_T \text{ AND } (0, 2, 4)_T \text{ then } (2, 3, 4)_T$$

with the fuzzy mean 3 as the consequence. Suppose that the first premise is fuzzy $\tilde{a}_1 = (2, 3, 4)_T$ and the second is crisp $a_2 = 3$.

If one applies the method to obtain a crisp consequence, the DOFs have to be calculated for each rule. Using the product inference method, *the DOF of the first rule is $\nu_1 = 0.5 \cdot 1 = 0.5$, and for the second rule, $\nu_2 = 1 \cdot 0.5 = 0.5$. Thus, by symmetry, the maximum and additive combination methods all yield the response $b = 2$ whether using the mean or the median defuzzification procedure.*

The fuzzy consequence is calculated using different level sets. One finds that the consequence here is not a triangular fuzzy number. For any $2 \leq a_1 \leq 3$ the fulfillment grade of rule one is $\nu_1 = (3 - a_1) \cdot 1$. For rule two, it is $\nu_2 = (a_1 - 2) \cdot 0.5$, from which the consequence is

$$\frac{(3 - a_1) \cdot 1 + (a_1 - 2) \cdot 0.5 \cdot 3}{(3 - a_1) + (a_1 - 2) \cdot 0.5} = \frac{a_1}{4 - a_1}$$

The corresponding membership value is $a_1 - 2$. This now has to be transformed into a proper membership function. Let

$$x = \frac{a_1}{4 - a_1}$$

and thus $1 \leq x \leq 3$ with

$$a_1 = \frac{4x}{1 + x}$$

The membership function of X defined on the interval $[1, 3]$ is, after some calculations, found to be:

$$\mu_{\tilde{b}}(x) = \frac{2x - 2}{x + 1} \quad \text{if } 1 \leq x \leq 3$$

If $3 < a_1 < 4$ then $\nu_1 = 0$ and the response always equals 3. Figure 3.9 shows the membership function of the fuzzy response.

The fuzzy mean of the fuzzy response $\tilde{b}$ is, in this case, 2.259 which is different from the consequence $b = 2$ obtained using the first method.

Note that the method of obtaining a fuzzy output is in fact the application of the extension principle to the rule system. In general it can be seen that this method to calculate the consequence requires much more effort than the case where a crisp consequence is to be determined. We are thus advocating to model systems that have a fuzzy premise $\tilde{a}$ with a crisp output whenever possible.

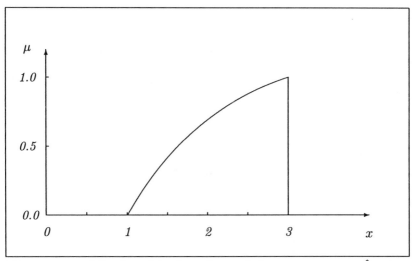

Figure 3.9. Membership functions of the consequence $\hat{b}$ using fuzzy premises in the rule system.

3.5   Rules with multiple responses

There are several cases where rules have to answer more than one question at the same time. It is often desirable to construct rules that do this at the same time — with rules having more than one consequence. There are two different possibilities to handle such problems:

1. To treat the different responses independently, which means that the rules are treated as a conjunction of as many independent rules as there are answers.

2. To treat the rule responses as fuzzy sets defined in the Cartesian product of individual response sets.

The first case is treated with the above described methods. In the second case, the DOF-s can be calculated as before. The combination methods do not need major modifications and can also be applied to multidimensional fuzzy sets too. The defuzzification methods require some additional definitions. The fuzzy mean (vector) of a multidimensional fuzzy set can be defined with a simple modification of the one dimensional definition, namely:

Definition 3.16 *The fuzzy mean of the fuzzy set B in the N dimensional Euclidean space is the vector $\mathbf{M}(B) = (M(B)_1, \ldots, M(B)_N)$*

where

$$M(B)_n = \frac{\int\limits_{-\infty}^{+\infty} \cdots \int\limits_{-\infty}^{+\infty} t_n \mu_B(t_1, \ldots, t_N)\, dt_1 \ldots dt_N}{\int\limits_{-\infty}^{+\infty} \cdots \int\limits_{-\infty}^{+\infty} \mu_B(t_1, \ldots, t_N)\, dt_1 \ldots dt_N} \qquad (3.5.1)$$

There is no simple generalization of the median for N dimensional spaces. Using the above fuzzy mean the combined responses can be defuzzified in the case of numerical responses.

Rule systems

In the previous chapter fuzzy rules were introduced and their combination and the defuzzification of the resulting responses was discussed. In this chapter we develop the different properties of rule answers depending on the set of rules and the methods of combination and defuzzification used.

This chapter provides a theoretical foundation for using fuzzy rule-based modeling, first by defining a rule system and its properties, and then by stating the necessary and sufficient conditions for a set of rules to encompass all possible "if ... then ..." statements. Finally, the selection of a resulting membership function is also discussed.

A set of rules operating on the same premises and having responses from the same set forms a rule system.

Definition 4.1 *The rule system $\mathcal{R}$ is the set of rules*

$$\text{If } A_{i,1} \odot A_{i,2} \odot \ldots \odot A_{i,K} \text{ then } B_i \qquad (4.0.2)$$

for $i = 1, \ldots, I$, where $A_{i,k}$-s are fuzzy subsets of X_k and B_i-s are fuzzy subsets of Y.

From this definition it seems that all arguments are used for every rule. However in practice this is not always the case. This problem can simply be overcome by taking $\mu_{A_{i,k}}(x) = 1$ for all $x \in X_k$ for arguments k which are not used in rule i. This way all rules are formulated with a common set of arguments.

A special case of rule systems are those where both the arguments and the responses are fuzzy numbers. These rule systems are the so-called numerical rule systems:

Definition 4.2 *A rule system $\mathcal{R}$ is a numerical rule system if for every i and k both $A_{i,k}$ and B_i are fuzzy numbers.*

Obviously more general statements can be formulated on numerical rule systems than on arbitrary ones. Linguistic variables can often be used to transform non-numerical rule premises and responses into a numerical form.

In order to calculate the response of a rule system, the DOF-s have to be evaluated using a selected inference method. In the next step the responses of the individual rules are combined with the help of a combination method. Finally a defuzzification method transforms the fuzzy rule response to a crisp result. This procedure can be considered as the definition of a function assigning the final (crisp) result to the input. In this chapter the properties of such a function are also investigated.

4.1   Completeness and redundancy

For a given rule system $\mathcal{R}$ two natural questions arise:

1. Are there enough rules to cover all possible cases that may occur in both the calibration and validation phases of the model?

2. Are there unnecessary rules in the system which might then be removed and if so, which ones?

The answers to these questions depend both on the combination and defuzzification methods used. The most important cases are treated in this section.

4.1.1   Completeness

A rule system is complete (or "consistent") on the set $\mathcal{A}$ if it can provide an answer to all possible "questions" concerning the phenomenon modeled. Formally:

Definition 4.3 *A rule system is complete if for every premise vector* $(a_1, \ldots, a_K) \in \mathcal{A}$ *the corresponding (combined) response set* $B(a_1, \ldots, a_K)$ *is a nonempty fuzzy set.*

Note that the definition requires not only that for every argument

$$(a_1, \ldots, a_K) \in \mathcal{A}$$

there should be a rule i with non-empty response, but it also requires the combined response to be non-empty.

The completeness of a rule system depends on the set $\mathcal{A}$ on which the rule responses are to be found, and on the combination method used. The defuzzification method does not influence the completeness of a rule system. As already pointed out in the discussion of the combination methods, the minimum combination seeks the difficult task of finding complete agreement, while the other methods only exclude responses that are impossible for each rule. Therefore, it is not surprising that the condition of completeness is easier to fulfill using the maximum or additive combination methods than using the minimum one. The following proposition can be formulated:

Proposition 4.1 *A rule system used with maximum or additive combination methods is complete on the set A if and only if for each premise vector $(a_1, \ldots, a_K) \in A$ there is a rule i such that $D_i(a_1, \ldots, a_K) > 0$*

The proof of this proposition is trivial. Note that the rule responses do not influence the completeness of the rule system if maximum or additive combination methods are used. However, this is not true for the minimum combination. Thus, the conditions for completeness are more complicated. Only the case of rules with fuzzy numbers as consequences is considered here.

Proposition 4.2 *A rule system $\mathcal{R}$ with a fuzzy number consequence set, and being used with the minimum combination is complete if and only if*

1. *for each $(a_1, \ldots, a_K) \in A$ there is a rule i such that $D_i(a_1, \ldots, a_K) > 0$ and*

2. *For any two rules i and j if there is an $(a_1, \ldots, a_K) \in A$ such that the degrees of fulfillment $D_i(a_1, \ldots, a_K) > 0$ and $D_j(a_1, \ldots, a_K) > 0$ then $B_i \cap B_j \neq \emptyset$ (there is such an x that $\min(\mu_{B_i}(x), \mu_{B_j}(x)) \neq 0)$.*

The proof of this proposition is not trivial and is thus given in the Appendix. The first condition is the same as for the maximum and additive combination methods. The second condition implies that for any number of rules with a positive DOF (for the same premises vector) the intersection of the responses is non-empty. This only holds if the responses are fuzzy numbers. The second condition can also be checked simply for any rule system. In contrast to the previous proposition where only the arguments need to be used to decide about the completeness of a rule system, the responses also have to be taken into account for the minimum combinations. Obviously if a rule system is not complete for the maximum or for the additive combination methods then it cannot be complete for the minimum combinations.

For rule systems with non- fuzzy numbers as consequences, there is no general condition which ensures the completeness of a rule system with the minimum combination. Note that the selected defuzzification procedure is irrelevant to the completeness of a rule system. Furthermore, the completeness of a rule system depends on the existence and the extent of overlap between the premises, which is thus a characteristic property of that system.

Example 4.1 *Consider a rule system consisting of the following two rules:*

$$\text{If } A_{1,1} = (1,2,3)_T \ \text{ then } B_1 = (1,2,3)_T$$

$$\text{If } A_{2,1} = (2,3,4)_T \ \text{ then } B_2 = (3,4,5)_T$$

This rule system is not complete on the interval $[2,3]$ *if the minimum combination is used, since*

$$\min \left(\mu_{B_1}(x), \mu_{B_2}(x) \right) = 0$$

for all $2 \leq x \leq 3$. *In contrast it is complete for the maximum and additive combination methods.*

The completeness of a rule system used with the maximum or additive combinations depends only on the "range" on which the arguments are defined. This "range", which is called the support of a rule, is defined as follows:

Definition 4.4 *The support of the rule*

$$\text{If } A_{i,1} \odot A_{i,2} \odot \ldots \odot A_{i,K} \text{ then } B_i$$

is the K *dimensional set:*

$$supp(i) = supp(A_{i,1}) \times \ldots \times supp(A_{i,K})$$

Proposition 4.1 can be reformulated using the definition of rule support as:

Proposition 4.3 *A rule system* $\mathcal{R}$ *used with the maximum or additive combinations is complete if and only if*

$$\mathcal{A} \subset \bigcup_{i=1}^{I} supp(i)$$

The proof of this proposition is evident by applying the definitions. The advantage of this formulation is that it can be checked without calculating any DOF-s. This condition is only a necessary condition for the minimum combination.

Figure 4.1 shows the supports of 4 rules (with two arguments) which form a complete rule system on a selected area. In contrast, Figure 4.2 shows a case where this is not true, as part of the area inside the dashed rectangle is not covered by any of the 4 rules.

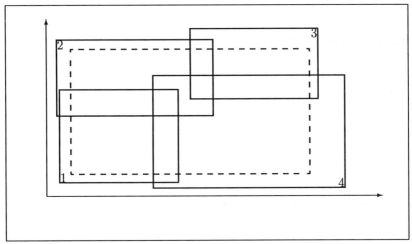

Figure 4.1.   Supports of 4 rules forming a complete rule system on the dashed area.

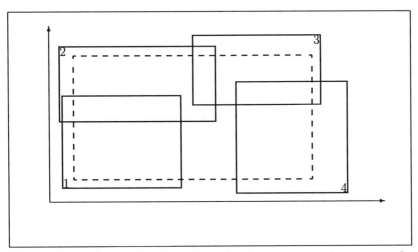

Figure 4.2.   Supports of 4 rules which are incomplete as a rule system on the dashed area.

Of course one can also ask the opposite question: which are the cases where the rule system $\mathcal{R}$ provides an answer? For this problem one can formulate the following proposition:

Proposition 4.4 *A rule system* $\mathcal{R}$ *used with the maximum or additive combinations is complete on the set*

$$\mathcal{A} = \bigcup_{i=1}^{I} supp(i)$$

Usually a rule system can answer more "questions" if it consists of more rules with possibly larger supports. However the formulation of the rules is not always obvious. Therefore, it is necessary to know which rules are really needed.

4.1.2 Redundancy

The overlap between the premises is an important property of fuzzy rule systems. It thus seems that rule systems may be at least partly redundant. However this is not true: even rules whose premises are completely covered by other rules might still modify the consequences.

Example 4.2 *Consider a rule system consisting of the following three rules:*

$$If\,(1,2,3)_T \quad then\,(0,2,4)_T$$

$$If\,(2,3,4)_T \quad then\,(3,5,7)_T$$

$$If\,(1.5,2.5,3.5)_T \quad then\,(1,3,5)_T$$

The premise of the third rule is fully covered by the premises of the other two rules. The rule system would remain complete (even for the minimum combination) on $[1.5, 2.5]$ *if the third rule were removed. The consequence using mean defuzzification and additive combination, with* $a_1 = 2.5$ *leads to*

$$\frac{2(0.5) + 5(0.5) + 1(3)}{0.5 + 0.5 + 1} = \frac{6.5}{2} = 3.25$$

removing the third rule gives a different result:

$$\frac{2(0.5) + 5(0.5)}{0.5 + 0.5} = \frac{3.5}{1} = 3.5$$

The possibility of removing a rule from a rule system without a major change in the consequence cannot be judged on the basis of only the premises. It should be done by comparing the response set. This problem will be reexamined in the last section of this chapter.

In order to know which rules have a chance to be removed, a measure of the overlap can is defined as:

Definition 4.5 *The overlap function O_v of a rule system $\mathcal{R}$ is defined as:*

$$O_v(a_1,\ldots,a_K) = |\{i; D_i(a_1,\ldots,a_K) > 0\}| = \sum_{i=1}^{I} \mathbf{1}\{supp(i)\}(a_1,\ldots,a_K)$$

where $|.|$ denotes the number of elements (cardinality) of a finite set, $\mathbf{1}\{supp(i)\}(a_1,\ldots,a_K)$ is the indicator function of the support of rule i, equal to 1 if $(a_1,\ldots,a_K)$ belongs to the support and 0 otherwise.

By definition the overlap function can only take on integer values, independently of the combination method, as it is defined on the premises.

Proposition 4.5 *A rule i can be removed from the rule system $\mathcal{R}$ without influencing its completeness if and only if*

$$\min\{O_v(a_1,\ldots,a_K) \,;\, (a_1,\ldots,a_K) \in supp(i) \cap \mathcal{A}\} > 1$$

The proof of this proposition is trivial. The proposition is valid for any combination method — even for the minimum combination. As pointed out already in Example 4.2, the removal of a rule can substantially influence the responses.

If the overlap function equals 1 in a neighborhood, then the rule response is constant for non-cresting combination methods in that neighborhood independently of the defuzzification methods selected, as only one rule applies. The constant value of the response function itself depends on the combination and defuzzification methods. Cresting can result in different values depending the DOF of the rule. As an illustration of this assertion, consider the following example:

Example 4.3 *Suppose that only the rule*

$$If \, (0,2,3)_T \quad then \, (0,1,4)_T$$

applies in the interval $[1,2]$. The minimum and maximum combinations yield to the same consequence set on this interval. The h level set of *the response is* $[h, \frac{4-h}{3}]$. *Using the formula for the calculation of the fuzzy mean for piecewise linear membership function Eq. 2.2.9 one can calculate the mean of the crested response* $B_1(h)$ *as:*

$$M(B_1(h)) = \frac{h\frac{2h+0}{6}h + \left(\frac{4-h}{3} - h\right)\left(\frac{\frac{8-2h}{3}+h}{6}h + \frac{\frac{4-h}{3}+2h}{6}h\right)}{h\frac{h}{2} + \left(\frac{4-h}{3} - h\right)h + \left(4 - \frac{4-h}{3}\right)\frac{h}{2}} +$$

$$+ \frac{\left(4 - \frac{4-h}{3}\right)\left(\frac{4+\frac{8-2h}{3}}{6}h\right)}{h\frac{h}{2} + \left(\frac{4-h}{3} - h\right)h + \left(4 - \frac{4-h}{3}\right)\frac{h}{2}}$$

After some algebra, one can simplify the above formula yielding:

$$M(B_1(h)) = \frac{-2h^2 - 5h + 52}{9(4 - h)}$$

For an $x \in [1, 2]$ *the corresponding membership level is* $h = \frac{x}{2}$, *thus the response with the cresting combination leads to:*

$$R(x) = \frac{-x^2 - 5x + 104}{9(8 - x)}$$

which is not constant on the interval $[1, 2]$. *This property that the combination is not constant even if only one rule applies is a major difference between the cresting and non-cresting combination methods.*

4.2   Variables to be used for rule systems

One of the most important questions in modeling is which variables to take into account. This depends on the decision of the modeler. If a rule system delivers similar responses, and/or similar agreement with the available data or knowledge after including an additional variable then this variable can be omitted since it does not improve system performance.

4.2.1   Hybrid rule systems

In the case when the effect of a number of variables is explicitly known and can be described with the help of simple functions (usually linear ones) it is not necessary to include them into the rule system. Instead one can construct a hybrid system consisting of explicit functional part(s)

$f(a_1, \ldots, a_l)$ and a rule system on the remaining variables $a_{l+1}, \ldots, a_I$. The multiplicative form is:

$$R(a_1, \ldots, a_I) = f(a_1, \ldots, a_l) R(a_{l+1}, \ldots, a_I) \qquad (4.2.1)$$

The additive form is:

$$R(a_1, \ldots, a_I) = f(a_1, \ldots, a_l) + R(a_{l+1}, \ldots, a_I) \qquad (4.2.2)$$

other combination forms can also be used. On the other hand, it will be shown later in this Chapter that any continuous function $f(a_1, \ldots, a_K)$ can be approximated by a set of fuzzy rules. One may thus select whichever modeling approach is simpler and clearer.

4.2.2   Interaction of variables

Rule systems can be formulated with or without considering interactions between the variables used as rule arguments. Non-interacting variables lead to rule systems of the form:

$$\text{If } A_{i,1} \text{ then } B_i$$

$$\text{If } A_{j,2} \text{ then } B_j$$

As mentioned above, all these rules can also be formulated using all arguments by using the entire set X_k as a fuzzy argument $A_{j,k}$ if a_k was not used as variable in rule j.

Partial interactions can also be taken into account. Which arguments interact and which do not is a decision to be taken during the assessment of a rule system. The fewer interactions that are considered, the fewer rules thta have to be used.

In fact one of the most important problems in using rule-based systems is the identification of the variables to be used, and the selection of the possible interactions between them. This task requires the knowledge of the system to be modeled. It is a major difference in comparison to neural nets, where only the variables are to be identified. The modelers knowledge is in the fuzzy rule- based case taken into account, which often yields robust models.

4.3 Rules and continuous functions

A rule system with a combination method and a defuzzification procedure produces a function, where the dependent variables are the actual values of the arguments. It is interesting to see the possible answers of a rule system as a function of its arguments. For this task only "really fuzzy" rule systems are considered:

Definition 4.6 *A numerical rule system $\mathcal{R}$ is non-degenerate if for every rule i, every premise $A_{i,k}$ and every rule response B_i the membership functions $\mu_{A_{i,k}}$ are continuous.*

In the classical (binary valued logic) case, the arguments $A_{i,k}$ are proper sets. The associated membership functions are the indicator functions, which are not continuous; thus a classical rule system is "degenerate". Non-degenerate rule systems are fuzzy in all their arguments. A numerical rule system with only non-crisp arguments is non-degenerate. This definition ensures that the DOF of a rule is a continuous function of the premises.

4.3.1 Continuity of rule system response functions

Using the previous definitions the first proposition of the response function can be formulated as follows:

Proposition 4.6 *If $\mathcal{A} = [a_1^-, a_1^+] \times \ldots \times [a_K^-, a_K^+]$ $\mathcal{R}$ is a non-degenerate complete numerical rule system on $\mathcal{A}$ and $R(a_1, \ldots, a_K)$ is obtained using the mean defuzzification of the response set $B(a_1, \ldots, a_K)$, then it is a continuous function on $\mathcal{A}$.*

The maximum and the median defuzzifications do not necessarily deliver continuous functions as responses, because of possible discontinuities and non-uniqueness of the maximum and median as discussed in Chapter 2. This is an important argument for the use of the mean defuzzification for rule systems with rule responses defined on a continuous scale.

The rule response functions are not necessarily piecewise linear functions. Depending on the underlying rules, they can have strongly nonlinear shapes:

Example 4.4 *Consider the rule system consisting of the following three rules:*

$$\text{If } (1, 2, 3)_T \quad \text{then } (0, 1, 2)_T$$

$$If\ (3, 4, 5)_T\ \ then\ (2, 3, 4)_T$$

$$If\ (0, 3, 6)_T\ \ then\ (0, 2, 4)_T$$

This is a complete and non-degenerate rule system on the interval $[1, 5]$. *Considering the normed weighted sum combination method with mean defuzzification, the function resulting from this rule system can be derived as follows:*
For $1 \le x \le 2$ *the DOFs are* $(x - 1), 0$ *and* $\frac{x}{3}$ *respectively. Thus the consequence is:*

$$\frac{1(x - 1) + 3(0) + 2(\frac{x}{3})}{(x - 1) + 0 + \frac{x}{3}} = \frac{5x - 3}{4x - 3}$$

For $2 \le x \le 3$ *the DOFs are* $3 - x, 0$ *and* $\frac{x}{3}$, *respectively. Thus the consequence is:*

$$\frac{1(3 - x) + 3(0) + 2(\frac{x}{3})}{(3 - x) + 0 + \frac{x}{3}} = \frac{9 - x}{9 - 2x}$$

For $3 \le x \le 4$ *the DOFs are* $0, x - 3$ *and* $\frac{6-x}{3}$, *respectively. Thus the consequence is:*

$$\frac{1(0) + 3(x - 3) + 2(\frac{6-x}{3})}{0 + x - 3 + \frac{6-x}{3}} = \frac{7x - 15}{2x - 3}$$

For $4 \le x \le 5$ *the DOFs are* $0, 4 - x$ *and* $\frac{6-x}{3}$, *respectively. Thus the consequence is:*

$$\frac{1(0) + 3(5 - x) + 2(\frac{6-x}{3})}{0 + 5 - x + \frac{6-x}{3}} = \frac{57 - 11x}{21 - 4x}$$

The function so defined is continuous as represented in Figure 4.3. Note that three crisp rules would only have given three different values as consequences. Minimum or maximum combination methods lead to different non-linear functions.

Thus, by using fuzzy rules, objects in discrete categories can be represented by continuous indices: a "high, medium, low" forecast is now given a continuous representation; or else, the severity of a patient state has become a continuous variable, as illustrated in Chapter 10. The next

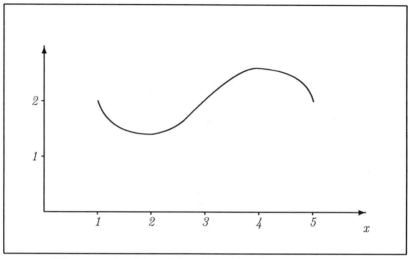

Figure 4.3.   **Resulting function in example 4.3**

proposition makes it possible to approximate any continuous functions by a set of fuzzy rules.

Proposition 4.7 *If $\mathcal{A} = [a_1^-, a_1^+] \times \ldots \times [a_K^-, a_K^+]$ and $f(a_1, \ldots, a_K)$ is a continuous function on $\mathcal{A}$ then for any $\varepsilon > 0$, any inference combination and any defuzzification method there is a rule system $\mathcal{R}$ such that*

$$|f(a_1, \ldots, a_K) - R(a_1, \ldots, a_K)| < \varepsilon$$

for each $(a_1, \ldots, a_K)$.

Classical rule systems define their conclusions on the premise sets. Therefore the number of possible consequences cannot exceed the number of rules and the resulting function is (unless all consequences are equal) not continuous. Besides the comfortable formulation of the rules this is a further advantage of fuzzy rule-based models especially in control problem. Similar results have been found by Wang and Mendel (1990). Another approach for approximating a function f by a set of rules has been given in Dubois and Prade (1994b) and Dubois et al. (1995).

Combining propositions 4.6 and 4.7 one finds the following important corollary:

Corollary 4.1 *Every non-degenerate rule system, inference and combination method with mean defuzzification can be replaced by a rule system* *using the AND operator, the product inference, weighted combination and fuzzy mean defuzzification.*

Obviously this corollary does not mean that the rule response functions corresponding to different combination methods are identical. As already pointed out in Chapter 3, the various combination methods usually lead to different responses. This corollary means that the additional effort required by the minimum and maximum combination and the corresponding unpleasant calculation of the fuzzy mean is not worth the effort. Instead the much simpler additive combination methods can be used with similar results. The only reason for using other methods may be specific to the case being investigated, in particular because of a simpler formulation of the rules. Unfortunately there is no similar result for rule systems with discrete fuzzy response sets. In Chapter 7 these systems will be discussed in detail. For all other illustrative examples described in the subsequent chapters, we use additive combination and mean defuzzification.

To compare the rule response functions corresponding to the same rule system used with different combination methods, consider the following example:

Example 4.5 *Consider the rule system consisting of the following four rules:*

$$\text{If } (1,2,3)_T \quad \text{then } (0,1,2)_T$$

$$\text{If } (2,3,4)_T \quad \text{then } (2,3,4)_T$$

$$\text{If } (3,4,5)_T \quad \text{then } (4,5,7)_T$$

$$\text{If } (4,5,6)_T \quad \text{then } (0,2,4)_T$$

This is a non-degenerate rule system on the interval $[2,5]$. It is complete for the maximum and additive combination methods. In contrast in the case of the minimum combination for all x values in the interval $[2,5]$ the minimum of the response membership functions is 0. Thus this rule system is not complete for the minimum combination. As the responses of the rules with positive DOF-s are non-overlapping for every a value in the interval $[2,5]$ the weighted sum and the maximum combinations give the same results. Figure 4.4 shows the rule response functions corresponding to the different combination methods, obtained using the fuzzy mean as defuzzification. Note that there is no big difference between

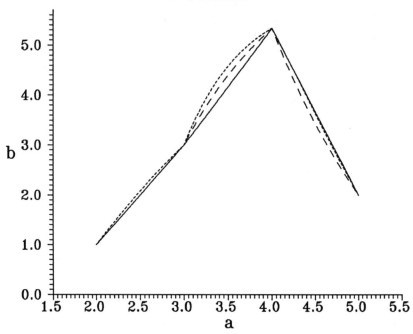

Figure 4.4.   Rule response function for Example 4.6 using the first rule responses, solid = normed weighted sum, dashed (long dashes) = weighted sum and maximum, dashed (short dashes) = cresting maximum combination method.

the three functions. Changing the rule responses to overlapping ones by defining the new rule system:

$$\text{If } (1,2,3)_T \quad \text{then } (0,1,4)_T$$

$$\text{If } (2,3,4)_T \quad \text{then } (2,3,4)_T$$

$$\text{If } (3,4,5)_T \quad \text{then } (3,6,7)_T$$

$$\text{If } (4,5,6)_T \quad \text{then } (1,2,4)_T$$

is considered. In this case the rule system is complete for the minimum combinations, too. Figure 4.5 shows the response functions. Note that in this case the differences are larger than before and the maximum and weighted sum combination responses are noticeably different. The minimum combination also delivers a response function that is nearly identi-

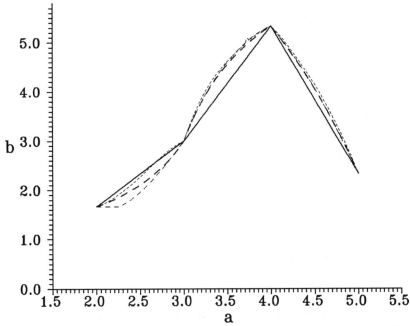

Figure 4.5. Rule response function for Example 4.6 using the second rule responses, solid = normed weighted sum, dashed = weighted sum, fine long dashes = maximum combination method, fine short dashes = cresting maximum combination method.

cal to that of the maximum combination. These examples also show that the response functions are quite robust with respect to the choice of the combination method.

The two propositions in this section show the property that most of the numerical rule systems deliver continuous response functions, and the converse property that all continuous functions can be closely approximated by rule systems. The natural question arises "Why should we use rule systems instead of continuous functions?" Traditionally the approach describing a relationship is to try to develop functions in a closed form. A reason for doing this is their simpler handling: as long as solutions had to be found "manually" there was no possibility to use functions whose description was not in a closed form. However the extensive use of computers during the past several years make a non-closed form description possible. The closed form functions selected have a specific shape as in linear regression, and their parameters are assessed with some kind of technique (for example, least squares). In this case one forces a specific shape to approximate the true (observed) data. It does not necessarily follow that the achieved approximation is reasonable. Furthermore the parameters of the function do not necessarily have a clear physical meaning. In contrast, the fuzzy rules do always possess clear interpretations. A function with an analytical form does not necessarily mean that the shape of the function is really well known to the person using it. Even in the two-dimensional space, the shape of a third order polynomial is already very hard to imagine. In contrast, rules define the function in selected neighborhoods usually known to the modeler and thus remain transparent.

The definition of continuous functions with the help of fuzzy rules is a very robust procedure. An error in one rule response only influences the function on the support of that rule. This is not true for arbitrary closed form continuous functions. An error in the coefficients of a polynom is dependent on the interval on which it is defined and on the degree of the monome whose coefficient is inexact.

The form of the continuous function depends on the combination method. To illustrate this, consider the following example.

Example 4.6 *Let a rule system on* $\mathcal{A} = [0, 2] \times [0, 2]$ *consist of 15 rules. These 15 rules are listed in Table 4.1.*

Table 4.1.   The rule set of Example 4.6

i	$A_{1,i}$	$A_{2,i}$	B_i
1	$(-0.5, 0.0, 0.5)_T$	$(-1, 0, 1)_T$	$(-0.141, 0.000, 0.133)_T$
2	$(0.0, 0.5, 1.0)_T$	$(-1, 0, 1)_T$	$(-0.003, 0.125, 0.210)_T$
3	$(0.5, 1.0, 1.5)_T$	$(-1, 0, 1)_T$	$(0.165, 0.250, 0.469)_T$
4	$(1.0, 1.5, 2.0)_T$	$(-1, 0, 1)_T$	$(0.335, 0.375, 0.603)_T$
5	$(1.5, 2.0, 2.5)_T$	$(-1, 0, 1)_T$	$(0.287, 0.500, 0.727)_T$
6	$(-0.5, 0.0, 0.5)_T$	$(0, 1, 2)_T$	$(0.216, 0.250, 0.379)_T$
7	$(0.0, 0.5, 1.0)_T$	$(0, 1, 2)_T$	$(0.165, 0.375, 0.610)_T$
8	$(0.5, 1.0, 1.5)_T$	$(0, 1, 2)_T$	$(0.375, 0.500, 0.790)_T$
9	$(1.0, 1.5, 2.0)_T$	$(0, 1, 2)_T$	$(0.388, 0.625, 0.702)_T$
10	$(1.5, 2.0, 2.5)_T$	$(0, 1, 2)_T$	$(0.485, 0.750, 0.916)_T$
11	$(-0.5, 0.0, 0.5)_T$	$(1, 2, 3)_T$	$(0.345, 0.500, 0.742)_T$
12	$(0.0, 0.5, 1.0)_T$	$(1, 2, 3)_T$	$(1.227, 1.375, 1.661)_T$
13	$(0.5, 1.0, 1.5)_T$	$(1, 2, 3)_T$	$(1.481, 1.750, 2.000)_T$
14	$(1.0, 1.5, 2.0)_T$	$(1, 2, 3)_T$	$(1.338, 1.625, 1.949)_T$
15	$(1.5, 2.0, 2.5)_T$	$(1, 2, 3)_T$	$(0.847, 1.000, 1.170)_T$

Rule response functions were calculated for this rule system using normed weighted sum, weighted sum, maximum and cresting maximum combination. The rule system is not complete for the minimum combination since for $a_1 = 0.2$, $a_2 = 0.5$ both rules 1 and 6 have a DOF of 0.4 but the responses are disjoint $(\min(\mu_{B_1}(x), \mu_{B_6}(x)) = 0$. Thus this combination method cannot be applied on the entire domain $\mathcal{A}$. In each case the mean defuzzification method was used to obtain a crisp response. The four different rule response functions are shown in Figure 4.6.

Normed weighted sum   **Weighted sum**

Maximum   **Cresting maximum**

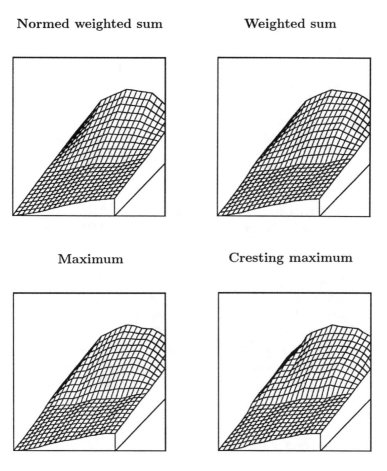

Figure 4.6. A comparison of the response functions corresponding to different combination methods in Example 4.7

One can see that the functions are very similar. The additive combinations provide very smooth surfaces while the maximum combination responses show small local irregularities.

To have a quantitative comparison, maximal, mean and mean absolute differences between the functions were calculated. Table 4.2 shows the differences between the response functions.

Table 4.2. Differences between the response functions in Example 4.6

Combination method	1	2	3	4
	Maximal difference			
1 Normed weighted sum	0.00	0.17	0.17	0.24
2 Weighted sum	0.17	0.00	0.09	0.13
3 Maximum	0.17	0.09	0.00	0.10
4 Cresting maximum	0.24	0.13	0.10	0.00
	Mean difference			
1 Normed weighted sum	0.00	-0.03	-0.04	-0.05
2 Weighted sum	0.03	0.00	-0.01	-0.02
3 Maximum	0.04	0.01	0.00	-0.01
4 Cresting maximum	0.05	0.02	0.01	0.00
	Mean absolute difference			
1 Normed weighted sum	0.00	0.03	0.04	0.05
2 Weighted sum	0.03	0.00	0.01	0.02
3 Maximum	0.04	0.01	0.00	0.02
4 Cresting maximum	0.05	0.02	0.02	0.00

Results of Table 4.2 show that the response surfaces obtained using the different combination methods do not differ substantially. Compared to the response function values, which are in between 0 and 1.75, the differences are insignificant. The normed weighted sum combination is only very slightly different from the others. This is due to the computational efficiency and the fact that this method gives greater weight to a crisper answer. We do not hesitate to recommend its use in modeling and control.

4.3.2 Smoothness of rule system response functions

As pointed out fuzzy rule systems produce continuous consequence functions which are a major advantage over traditional (binary valued) rule systems. The continuity of rule response functions only depends on the continuity of the membership functions of the arguments. An appropriate choice of the membership functions of the premises may be sufficient to ensure the differentiability of the response function. For example, the following proposition may be used:

Proposition 4.8 *Suppose that all rules of a non-degenerate numerical rule system are formulated with arguments $A_{i,k}$ which have differentiable membership functions $\mu_{A_{i,k}}(x)$ for all x real values: then using the mean defuzzification method the resulting function $R(a_1, \ldots, a_K)$ is differentiable.*

An example of such membership functions is that of L-R fuzzy numbers with the L-R functions:

$$L(x) = R(x) = \frac{1}{2}(\cos(\pi x) + 1)$$

Figure 2.9 in Chapter 2 shows the graph of such membership functions. In the case of premises with triangular or trapezoidal membership functions the resulting functions are usually not differentiable everywhere: the function defined in Example 4.4 is thus not differentiable at $x = 2$.

4.4 Membership functions in rule systems

Fundamental questions that arise are: What is a good rule? How crisp should arguments and responses be? What shape should the membership functions have? The answer to these questions is not obvious and cannot be found by considering rules as individual objects, but only as elements of a complete rule system.

As the DOF and the combination all depend on the selection of the various membership functions it is necessary to investigate the problem of their selection. We first examine the choice of response membership function

4.4.1 Response membership functions

The shape of the membership function of the rule responses determines the combined response. For the strongly recommended normed weighted sum combination method and the mean defuzzification, only the fuzzy means $M(B_i)$ of the responses B_i are used to obtain the defuzzified response function $R(a_1, \ldots, a_I)$. Thus, for this suggested method, the choice of the membership function is not relevant.

For the weighted sum combination, in addition to the fuzzy mean of the response, the area under the membership function is also needed. As pointed out in Chapter 3, rules with a larger area under the membership function have a greater influence on the combined and defuzzified result.

For the minimum combination methods, in order to have a complete rule system, it is preferable to have fuzzier responses.

4.4.2 Argument membership functions

Considering the arguments used in fuzzy rules the first question is how fuzzy should these arguments be? Taking arguments with very crisp (narrow) supports would necessarily lead to rule systems with many rules whereas very wide supports might lead to non-specific responses. The completeness of the rule system restricts the choice, as the supports have to cover the whole area. The response function is not a step function (being constant over in selected areas) if there is an overlap between the rule supports. Furthermore the sensitivity to error is smaller if there is a considerable overlap.

Once the support of the argument is specified, the next question is the selection of the exact shape of the membership function. Triangular or trapezoidal fuzzy numbers are the simplest candidates and are most often used. It is surprising to observe how small the difference is resulting from the exchange of a triangular membership function into an L-R fuzzy number. To illustrate this assertion, consider the following example.

Example 4.7 *Consider the rule system of Example 4.6 described in Table 4.1. Let triangular membership functions be replaced by L-R fuzzy numbers with $L(x) = R(x) = \frac{1}{2}\left(\cos(\pi x) + 1\right)$.*

Table 4.3. Differences between the response functions using triangular versus L-R fuzzy numbers with $L(x) = R(x) = \frac{1}{2}\left(\cos(\pi x) + 1\right)$ in Example 4.7

Combination method	Maximal	Mean	Mean absolute
Normed weighted sum	0.18	0.00	0.04
Weighted sum	0.17	0.01	0.04
Maximum	0.19	0.01	0.04
Cresting maximum	0.20	0.00	0.04

Table 4.3 shows the differences between the response surfaces corresponding to the triangular and L-R fuzzy number arguments and Figure

4.6 provides a sketch of these response surfaces. One can see that the differences are comparable to the differences between the combination methods.

The argument membership functions influence the differentiability of the response surface as stated in Proposition 4.8.

4.5 Sensitivity of the response functions

It is interesting to know how a rule system response function changes in the case of small changes of the rule responses (incorrectly specified rules). As already pointed out, rule system response functions are continuous in their arguments. It can be shown that they are also continuous functions of the responses in the fuzzy sense (Dubois and Prade, 1980a) for most combination and defuzzification methods. The exact statement and its proof are not given in this book, only the suggested case the normed weighted combination method with the fuzzy mean defuzzification is treated here. In this case, only the fuzzy mean of the rule responses is required. Thus, the rule response function can be regarded as a function of the individual rule responses $M(B_1), \ldots, M(B_I)$. Suppose that rule response i is altered by ε_i giving the new response $M(B_i) + \varepsilon_i$ to rule i. Let $\nu_i = D_i(a_1, \ldots, a_K)$ be the DOF corresponding to $(a_1, \ldots, a_K)$ for rule i. In this case, the rule response function of the original rule system is:

$$R(a_1, \ldots, a_K) = \frac{\sum_{i=1}^{I} \nu_i M(B_i)}{\sum_{i=1}^{I} \nu_i} \qquad (4.5.1)$$

The altered rule response function is:

$$R'(a_1, \ldots, a_K) = \frac{\sum_{i=1}^{I} \nu_i (M(B_i) + \varepsilon_i)}{\sum_{i=1}^{I} \nu_i} \qquad (4.5.2)$$

Thus the difference is

$$|R(a_1, \ldots, a_K) - R'(a_1, \ldots, a_K)| = \left| \frac{\sum_{i=1}^{I} \nu_i \varepsilon_i}{\sum_{i=1}^{I} \nu_i} \right| \leq \max_i(\varepsilon_i) \qquad (4.5.3)$$

This shows that the rule response functions are not sensitive to small errors in the individual rule responses. Furthermore, an error only influences the response function on its support. From Equation (4.5.3) it follows that if only a single rule, say l, is erroneous then the difference is:

$$|R(a_1, \ldots, a_K) - R'(a_1, \ldots, a_K)| = \left| \frac{\nu_l \varepsilon_l}{\sum_{i=1}^{I} \nu_i} \right| \qquad (4.5.4)$$

From this, one can conclude that the larger the overlap is on the support of rule l the smaller the change in the response function, as the denominator increases if more rules apply. This also points out the importance of the overlap in fuzzy rule systems.

5

Rule construction

In this chapter, possibly the most important part in the process of developing fuzzy rule based systems is discussed, namely, the assessment of rules. Assessment of rules is a procedure where knowledge and/or available data are translated or encoded into rules. Since rule system responses depend both on the combination and defuzzification method applied, this choice has to be taken into account in the assessment.

Fuzzy rule-based modeling is centered around the definition and verification of a rule system. At least three different ways to define a rule system may be distinguished:

1. The rules are known by the experts and can be defined directly.

2. The rules can be assessed by the experts directly, but available data should be used to update them.

3. The rules are not known explicitly, but the variables required for the description of the system can be specified by the experts.

4. Only a set of observations is available, and a rule system has to be constructed to describe the interconnections between the input/output elements of the data set.

Each of these approaches is now described in some detail. The order of the methods is altered for an easier understanding.

5.1 Explicit rule specification

To illustrate the concept of explicit rule use, consider a simple model of algae growth in a water body receiving a steady loading of nutrients including dissolved oxygen. It has been observed that available nutrients x and algae biomass y fluctuate. Let x and y be measured on the scale between 0 and 1. For the rules construction, let only two state descriptors high (H) and low (L) be used. The transition between the states can be described as:

$$HH \rightarrow LH \rightarrow LL \rightarrow HL \rightarrow HH \qquad (5.1.1)$$

(for example HL means $x = H$ and $y = L$)

Starting with HH if available nutrients and algae are H at time t algae reduce nutrients to L at time $t + 1$; in turn, because nutrients are insufficient, algae become L. The nutrients are replenished and become H; again algae become H in the next time period.

Let $\hat{H}$ and $\hat{L}$ be triangular fuzzy numbers defined on the unit interval:

$$\hat{H} = (0.4, 1.0, 1.0)_T \quad \hat{L} = (0.0, 0.0, 0.7)_T$$

with respective means given by Eq. (2.2.6) as: $M(\hat{H}) = \frac{2.4}{3} = 0.8$ and $M(\hat{L}) = \frac{0.7}{3} = 0.233$, and medians given by Eq. (2.2.14)as: $m(\hat{H}) = 0.4 + 0.3\sqrt{2}$ and $m(\hat{L}) = 1 - 0.35\sqrt{2}$.

The fuzzy rule-based algae model thus defined can be used to construct the state vector trajectory in a continuous space; given an initial state, this trajectory is unique.

Let the initial state be:

$$(x(0), y(0)) = (0.5, 0.6)$$

From the definition of $\hat{H}$ and $\hat{L}$ one finds the membership function values corresponding to this initial state. For nutrients x:

$$\mu_L(0.5) = \frac{2}{7} \quad \mu_H(0.5) = \frac{1}{6} \tag{5.1.2}$$

and for algae y:

$$\mu_L(0.6) = \frac{1}{7} \quad \mu_H(0.6) = \frac{2}{6} \tag{5.1.3}$$

Since the state is defined by an AND rule, the fulfillment grade $\nu(i, j)$ of a rule, with $(i, j) \in \{L, H\}^2$, is the product:

$$\nu(i, j) = \mu_i(0.5)\mu_j(0.6) \tag{5.1.4}$$

Table 5.1 shows the degree of fulfillment (DOF) of every state. In order to calculate the next state, all the information in this table is used and may be displayed in the form of transition rules as in Table 5.2. The combination of fuzzy state descriptors is done next by using the fuzzy mean.

Table 5.1. Calculation of the degree of fulfillment using the multiplicative inference method

	$\mu_H(0.6) = \frac{2}{6}$	$\mu_L(0.6) = \frac{1}{7}$
$\mu_H(0.5) = \frac{1}{6}$	$\frac{1}{18}$	$\frac{1}{42}$
$\mu_L(0.5) = \frac{2}{7}$	$\frac{2}{21}$	$\frac{2}{49}$

Table 5.2. Transition table for an initial state $(x(0), y(0)) = (0.5, 0.6)$

Rule	DOF
$HH \rightarrow LH$	$\frac{1}{18}$
$LH \rightarrow LL$	$\frac{2}{21}$
$LL \rightarrow HL$	$\frac{2}{49}$
$HL \rightarrow HH$	$\frac{1}{42}$

The fuzzy mean combination of rules given by Eq. 3.3.4, in which $\mu_{B_i}(x)$ is replaced by its mean value, with $\beta_H = \frac{1}{0.3}$ and $\beta_L = \frac{1}{0.35}$, yields for the nutrients:

$$x(1) = \frac{\frac{1}{18}\frac{0.7}{3}0.35 + \frac{2}{21}\frac{0.7}{3}0.35 + \frac{2}{49}\frac{2.4}{3}0.30 + \frac{1}{42}\frac{2.4}{3}0.3}{\frac{1}{18}0.35 + \frac{2}{21}0.35 + \frac{2}{49}0.3 + \frac{1}{42}0.3} = 0.4319 \quad (5.1.5)$$

The fuzzy mean combination of rules with normed responses given by Eq. 3.3.5 yields for the nutrients:

$$x(1) = \frac{\frac{1}{18}\frac{0.7}{3} + \frac{2}{21}\frac{0.7}{3} + \frac{2}{49}\frac{2.4}{3} + \frac{1}{42}\frac{2.4}{3}}{\frac{1}{18} + \frac{2}{21} + \frac{2}{49} + \frac{1}{42}} = 0.4034 \quad (5.1.6)$$

For the algae the two corresponding results are:

$$y(1) = \frac{\frac{1}{18}\frac{2.4}{3}0.3 + \frac{2}{21}\frac{0.7}{3}0.35 + \frac{2}{49}\frac{0.7}{3}0.35 + \frac{1}{42}\frac{2.4}{3}0.3}{\frac{1}{18}0.3 + \frac{2}{21}0.35 + \frac{2}{49}0.35 + \frac{1}{42}0.3} = 0.4222 \quad (5.1.7)$$

and

$$y(1) = \frac{\frac{1}{18}\frac{2.4}{3} + \frac{2}{21}\frac{0.7}{3} + \frac{2}{49}\frac{0.7}{3} + \frac{1}{42}\frac{2.4}{3}}{\frac{1}{18} + \frac{2}{21} + \frac{2}{49} + \frac{1}{42}} = 0.4419 \quad (5.1.8)$$

If the min-max inference method is used, then the fulfillment grade $\nu(i,j)$ where i and j can take on values L or H, is the minimum of the row or column entry, as shown in Table 5.3.

$$\nu(i,j) = \min(\mu_i(0.5)\mu_j(0.6)) \qquad (5.1.9)$$

Table 5.3. Calculation of the degree of fulfillment using the min-max inference method

	$\mu_H(0.6) = \frac{2}{6}$	$\mu_L(0.6) = \frac{1}{7}$
$\mu_H(0.5) = \frac{1}{6}$	$\frac{1}{6}$	$\frac{1}{7}$
$\mu_L(0.5) = \frac{2}{7}$	$\frac{2}{7}$	$\frac{1}{7}$

The fuzzy mean combination of rules thus yields for the nutrients:

$$x(1) = \frac{\frac{1}{6}\frac{0.7}{3} + \frac{2}{7}\frac{0.7}{3} + \frac{1}{7}\frac{2.4}{3} + \frac{1}{7}\frac{2.4}{3}}{\frac{1}{7} + \frac{1}{7} + \frac{2}{7} + \frac{1}{6}} = 0.4526 \qquad (5.1.10)$$

and for the algae:

$$y(1) = \frac{\frac{1}{6}\frac{2.4}{3} + \frac{2}{7}\frac{0.7}{3} + \frac{1}{7}\frac{0.7}{3} + \frac{1}{7}\frac{2.4}{3}}{\frac{1}{7} + \frac{1}{7} + \frac{2}{7} + \frac{1}{6}} = 0.4710 \qquad (5.1.11)$$

Comparing Eq. 5.1.7 with 5.1.10, and then Eq. 5.1.8 with 5.1.11 shows that there is some difference between the results of the product and the min-max combination methods. A validation procedure may be used to assess which of the combination method provides the best description of reality.

The same procedure is used to obtain $(x(t+1), y(t+1))$ as a function of $(x(t), y(t))$ for $t = 1, 2, \ldots$. Note that the initial state transition table is used at every step to construct a table analog to Table 5.2 but now the state vector values are defined as continuous variables on the unit square. After 2 time periods, an oscillatory behavior of both x and y is observed which is exactly the same as the solution of the coupled differential equations model. Figure 5.1 shows the trajectories of $x(t)$ and $y(t)$ using the product inference method.

Figure 5.1. **Algae growth model results** $x(t)$ (*algae*) **(thin lines)** $y(t)$ (*nutrients*) **(thick lines).**

5.2 Deriving rule systems from datasets

Quite often, a data set containing all possible variables and the corresponding responses are available and a plausible explanation of the response is sought. Typical approaches to model this problem are regression techniques; however, as noted in Chapter 4, in a regression the form of the function delivering the response is prescribed. In most cases the relationship is assumed to be linear or linearized, which restricts the model applicability. If non-linearities are also considered, both analytical and computational aspects of the techniques become much more difficult (Bárdossy et al., 1993b).

Given that there is no closed functional form of the response function for a rule system, different techniques have to be applied. There are at least two possibilities to derive rules from observation data sets when the explanatory variables or premises are given, namely:

1. The rule system structure is known: the premises are specified and the corresponding rule responses have to be assessed using the available data.

2. The rule system structure is unknown, so that the important variables and the corresponding explicit rules have to be specified.

In both cases a trial and error approach can be used to find an appropriate rule system. This is unfortunately a very time consuming exercise even in the case of a few variables, hence recent developments in the use of artificial neural networks to assess an initial set of rules. (Kosko, 1992; Bárdossy et al., 1993c; Muster et al., 1994) In certain cases it is possible to derive rules directly from the data sets. In these cases the rules can be refined subsequently with a trial and error redefinition of the corresponding fuzzy sets or else, using neural nets again.

5.3   Known rule structure

5.3.1   The training set

Suppose that the set of possible outputs consists of all the relevant variables and the corresponding response. This set $\mathcal{T}$ is called the training or calibration set of observations:

$$\mathcal{T} = \{(a_1(s), \ldots, a_K(s), b(s)) \; ; s = 1, \ldots, S\} \qquad (5.3.1)$$

A rule system that delivers a response close to the observed one has to be assessed using this set $\mathcal{T}$. For this purpose, three techniques are developed below, namely,

1. the counting algorithm

2. the weighted counting algorithm

3. the least squares algorithm

The traditional training set consists of a large number of observation data. However, fuzzy rules can also be used for the purpose of simplifying complicated system models. In such a case, detailed model outputs can be used as a synthetic training set; this technique is illustrated in greater detail in Chapter 9, as well as in Bárdossy and Disse (1993). The advantage of this approach lies in the possibility of using arbitrarily large data sets.

5.3.2   The counting algorithm

The most straightforward method for rule assessment is to define the
rule premises explicitly and subsequently to define the response corresponding to a rule with the help of all observed data for which the rule
is applicable. Suppose that all variables including the response are all
numerical and defined on a continuous scale. The rules can now be assessed by defining the fuzzy set supports for the fuzzy numbers $A_{i,k}$, and
identifying the corresponding responses. The algorithm is

1. Define the support $(\alpha_{i,k}^-, \alpha_{i,k}^+)$ of the rule argument $A_{i,k}$.

2. The $A_{i,k}$ is assumed to be a triangular fuzzy number $(\alpha_{i,k}^-, \alpha_{i,k}^1, \alpha_{i,k}^+)_T$
   where $\alpha_{i,k}^1$ is the mean of all possible $a_k(s)$ values which fulfill at least
   partially the ith rule:

$$\alpha_{i,k}^1 = \frac{1}{N_i} \sum_{s \in R_i} a_k(s) \qquad (5.3.2)$$

Here R_i denotes the set of all those premise value vectors that fulfill
at least in part the ith rule, R_i forms a subset of the training set $\mathcal{T}$:

$$R_i = \{(a_1(s), \dots, a_K(s), b(s)) \in \mathcal{T} \; ; \; \text{such that}$$

$$a_k(s) \in (\alpha_{i,k}^-, \alpha_{i,k}^+) \text{ for } k = 1, \dots, K\} \qquad (5.3.3)$$

N_i is the number of elements in R_i.

3. The corresponding response is also assumed to be a triangular fuzzy
   number $(\beta_i^-, \beta_i^1, \beta_i^+)_T$ where β_i^- is the minimal "answer" corresponding to R_i:

$$\beta_i^- = \min_{s \in R_i} b(s) \qquad (5.3.4)$$

β_i^1 is the mean "answer" corresponding to R_i:

$$\beta_i^1 = \frac{1}{N_i} \sum_{s \in R_i} b(s) \qquad (5.3.5)$$

and β_i^+ is the maximal "answer" corresponding to R_i:

$$\beta_i^+ = \max_{s \in R_i} b(s) \qquad (5.3.6)$$

Thus the assessed rules are:

$$\text{IF } (\alpha_{i,1}^-, \alpha_{i,1}^1, \alpha_{i,1}^+)_T \text{ AND} \ldots (\alpha_{i,K}^-, \alpha_{i,K}^1, \alpha_{i,K}^+)_T \text{ THEN } (\beta_i^-, \beta_i^1, \beta_i^+)_T$$

This algorithm can be used in conjunction with all combination and defuzzification methods. Note that the definition of the supports $(\alpha_{i,k}^-, \alpha_{i,k}^+)$ plays a crucial role in the assessment of the rules. At this point a good knowledge of the system to be described is essential so that a representative rule system may be obtained. The rules thus assessed can be validated by comparing the response of the rule system for the elements of the data set with the "exact" responses, using whenever possible (enough data available) a split sampling procedure in which the training set is different from the test or validation set.

5.3.3   The weighted counting algorithm

The counting algorithm does not use the DOF of a rule to find the rule response. Observed responses corresponding to premises with very low DOF-s are considered with the same weight as those for which the rule is strongly applicable. Therefore it is reasonable to modify the above algorithm by taking into account the DOF values. Using the same notation as for the counting algorithm, the weighted counting algorithm can be described as follows:

1. Define the membership functions of the premises. This step can be performed in the same way as for the counting algorithm using Eq. 5.3.2.

2. Calculate the DOF-s $\nu_i(s)$ for each premise vector $(a_1(s), \ldots, a_K(s))$ corresponding to the training set $\mathcal{T}$ and each rule i whose premises were defined in Step 1.

3. Select a number $\varepsilon > 0$ such that only responses with a DOF of at least equal to ε will be considered in the construction of the rule response. The corresponding response is also assumed to be a triangular fuzzy number $(\beta_i^-, \beta_i^1, \beta_i^+)_T$ where β_i^+ is the minimal "answer" with a degree of fulfillment of at least ε

$$\beta_i^- = \min_{\nu_i(s) > \varepsilon} b(s) \qquad (5.3.7)$$

β_i^1 is the DOF weighted mean "answer"

$$\beta_i^1 = \frac{\sum_{\nu_i(s)>\varepsilon} \nu_i(s)b(s)}{\sum_{\nu_i(s)>\varepsilon} \nu_i(s)} \qquad (5.3.8)$$

and β_i^+ is the maximal "answer" with a degree of fulfillment of at least ε

$$\beta_i^+ = \max_{\nu_i(s)>\varepsilon} b(s) \qquad (5.3.9)$$

The value of ε has to be selected so that for each rule a sufficient number of elements of the training set is considered. The higher ε the fewer elements are used to define the responses, and the crisper are the assessed rules.

5.3.4   The least squares algorithm

As pointed out in the previous sections, rule systems produce continuous response functions. Therefore, one may use traditional methods of function fitting such as least squares techniques.

Suppose that all variables including the response are numerical and defined on a continuous scale. The rule system to be assessed should be used with the normed weighted sum combination method and the mean defuzzification. This is not a severe restriction as the response functions are not very sensitive to the selection of the combination and defuzzification method (see Chapters 3 and 4). In this case, as mentioned above, it is sufficient to define the fuzzy mean $M(B_i)$ of the rule responses. The rules are assessed by defining the left-hand sides of the rules first as in the previous algorithm. The responses are assumed to minimize the sum of the squared error resulting from use of the rules:

$$\sum_s [R(a_1(s), \ldots, a_K(s)) - b(s)]^2 \qquad (5.3.10)$$

As the left hand side of the rules is supposed to be known, the DOF $\nu_i(s) = D_i(a_1(s), \ldots, a_K(s))$ corresponding to $(a_1(s), \ldots, a_K(s))$ can be calculated for each rule i. Then the rule response for the normed weighted sum combination method can be written as:

$$R(a_1(s), \ldots, a_K(s)) = \frac{\sum_{i=1}^I \nu_i(s)M(B_i)}{\sum_{i=1}^I \nu_i(s)} \qquad (5.3.11)$$

From Eqs. 5.3.10 and 5.3.11, the objective function to be minimized is the squared difference between the rule responses and the observed responses corresponding to the elements of the training set.

$$\sum_s \left(\frac{\sum_{i=1}^I \nu_i(s) M(B_i)}{\sum_{i=1}^I \nu_i(s)} - b(s) \right)^2 \qquad (5.3.12)$$

The necessary and sufficient condition for a minimum of this quadratic function is that its derivative with respect to the unknown $M(B_j)$-s be 0. Thus for every index j

$$\sum_s \left(\frac{\sum_{i=1}^I \nu_i(s) M(B_i)}{\sum_{i=1}^I \nu_i(s)} - b(s) \right) \frac{\nu_j(s)}{\sum_{i=1}^I \nu_i(s)} = 0 \qquad (5.3.13)$$

Rearranging terms, one obtains the linear equation system

$$\sum_s \frac{\sum_{i=1}^I \nu_i(s) \nu_j(s) M(B_i)}{(\sum_{i=1}^I \nu_i(s))^2} = \sum_s \frac{\nu_j(s) b(s)}{\sum_{i=1}^I \nu_i(s)} \qquad (5.3.14)$$

There are I unknowns $M(B_1), \ldots, M(B_I)$ and I equations; thus, the above system usually provides the unknown quantities. Note that the shapes of the premise (premises or consequences) membership functions are taken into account through the DOF-s ($\nu_i(s)$) in the solution, but their differentiability is not required for calculating the solution.

The algorithm can be described as follows:

1. Define the membership functions of the premises. This step can be performed in the same way as for the counting algorithm.

2. Assess and solve the linear equation system (5.3.14).

3. The final rule system is:

$$\text{If } A_{i,1} \text{ AND } A_{i,2} \text{ AND} \ldots \text{AND } A_{i,K} \text{ then } M(B_i) \qquad (5.3.15)$$

Note that this algorithm only delivers the fuzzy mean of the responses, as only this quantity is needed for the normed weighted sum combination method used in connection with the mean defuzzification. In contrast the counting and the weighted counting algorithms can be used for any combination and defuzzification method.

A disadvantage of the algorithm is that it might deliver rule responses $M(B_i)$ which are not reasonable as an individual response to that single rule. This problem however can be overcome by a combination of the counting and least squares algorithm (see Section 5.4.3).

5.3.5 Comparison of the algorithms

All three algorithms generate responses for numerical rule systems. While the first two can be applied with any kind of combination and defuzzification methods, the least squares is only applicable to the normed weighted sum combination method and the mean defuzzification. The first two algorithms produce triangular fuzzy number outputs, thus allowing a better insight into the uncertainty associated with the rule responses.

The following example illustrates the difference between the three algorithms:

Example 5.1 *25 observation data $(a_1(s), b(s))$ are given in Table 5.4. A rule system for the interval $[0, 8]$ is to be found. The data show a highly variable and non-linear behavior, making the modeling task quite difficult.*

Table 5.4. The training set T

s	$a(s)$	$b(s)$	s	$a(s)$	$b(s)$
1	0.09	0.22	14	3.70	4.30
2	0.28	0.57	15	3.91	4.94
3	0.34	0.76	16	4.25	3.72
4	0.65	1.47	17	4.37	4.25
5	0.67	1.64	18	4.43	3.16
6	1.53	0.14	19	5.12	-6.24
7	1.66	-0.31	20	5.21	-8.01
8	1.90	-1.40	21	5.42	-10.15
9	2.35	-2.75	22	5.48	-6.48
10	2.95	-1.24	23	5.72	-7.23
11	3.30	1.17	24	5.97	-5.79
12	3.61	3.23	25	7.50	1.00
13	3.64	3.49			

A rule system consisting of 7 rules is sought. Without any further prior information the supports of the fuzzy premises $A_{1,i}$ are taken as intervals of equal length. The supports are selected so as to provide a regular cover of the whole interval — with two to three rules applicable to each $a(s)$ value.

The element with unit membership was generated by means of the counting algorithm — and was taken as the same for the other two algorithms as well. Table 5.5 shows the rules together with the responses obtained using the counting algorithm. Note that Rule 7 gives a crisp answer — as there is only one element in the training set for which this rule can be applied. The other rules all have responses with large supports — as an extreme, Rule 5 possesses an interval (-10.15,4.94) that covers the entire range of responses. This high uncertainty is responsible for the poor performance of the resulting rule system.

Table 5.5. The rule system obtained using the counting algorithm

i	$A_{1,i}$	B_i
1	$(-1.20, 0.75, 1.80)_T$	$(-0.31, 0.64, 1.64)_T$
2	$(0.00, 1.24, 3.00)_T$	$(-2.75, -0.09, 1.64)_T$
3	$(1.20, 2.86, 4.20)_T$	$(-2.75, 1.16, 4.94)_T$
4	$(2.40, 4.04, 5.40)_T$	$(-8.01, 1.16, 4.94)_T$
5	$(3.60, 4.68, 6.60)_T$	$(-10.15, -1.29, 4.94)_T$
6	$(4.80, 5.77, 7.80)_T$	$(-10.15, -6.13, 1.00)_T$
7	$(6.00, 7.50, 9.00)_T$	$(1.00, 1.00, 1.00)_T$

For the weighted counting algorithm $\varepsilon = 0.5$ was selected. This value ensures that for each rule a sufficient number of data is considered. Table 5.6 shows the rules together with the responses obtained using such a weighted counting algorithm. In comparison to the counting algorithm, the supports of the responses are smaller — crisper responses are obtained because of the restriction on the DOF. However the support of the response of Rule 5 covers again almost the total range of the observed response set.

Figure 5.2 shows the training set and the response function of the seven rules for the counting and the weighted counting algorithm using normed weighted combination and mean defuzzification. One can see from the figure that the weighted counting algorithm delivers a response function that is closer to the training set. The correlation coefficient between the observed and the model responses is 0.71 for the counting algorithm and 0.90 for the weighted counting algorithm. There are not enough data for using a split sampling technique.

Table 5.6.   The rule system obtained using the weighted counting algorithm

i	$A_{1,i}$	B_i
1	$(-1.20,\ 0.75,\ 1.80)_T$	$(0.22,\ 1.01,\ 1.64)_T$
2	$(0.00,\ 1.24,\ 3.00)_T$	$(-1.40,\ 0.20,\ 1.64)_T$
3	$(1.20,\ 2.86,\ 4.20)_T$	$(-2.75,\ -0.99,\ 1.17)_T$
4	$(2.40,\ 4.04,\ 5.40)_T$	$(1.17,\ 3.66,\ 4.94)_T$
5	$(3.60,\ 4.68,\ 6.60)_T$	$(-10.15,\ -2.70,\ 4.25)_T$
6	$(4.80,\ 5.77,\ 7.80)_T$	$(-10.15,\ -7.24,\ -5.79)_T$
7	$(6.00,\ 7.50,\ 9.00)_T$	$(1.00,\ 1.00,\ 1.00)_T$

Consider now the least squares algorithm. In order to obtain a smooth response function instead of triangles, L-R fuzzy numbers with

$$L(x) = R(x) = \frac{1}{2}(\cos(\pi x) + 1)$$

are selected as rule premises.

Table 5.7 shows the rules together with the fuzzy means of the responses estimated with the least squares algorithm. Figure 5.3 sketches the training set and the response function of the seven rules for the same algorithm. Additionally a traditional least squares, a polynomial fit of the data is shown. A sixth order polynomial selected as the number of parameters is the same as the rule responses determined by the least squares rule estimation.

Table 5.7.   The rule system obtained using the least squares algorithm

i	$A_{1,i}$	$M(B_i)$
1	$(-1.20, 0.75, 1.80)_{LR}$	0.80
2	$(0.00, 1.24, 3.00)_{LR}$	0.95
3	$(1.20, 2.86, 4.20)_{LR}$	-5.00
4	$(2.40, 4.04, 5.40)_{LR}$	9.01
5	$(3.60, 4.68, 6.60)_{LR}$	-6.10
6	$(4.80, 5.77, 7.80)_{LR}$	-8.21
7	$(6.00, 7.50, 9.00)_{LR}$	1.49

Note that the rule response function provides a better fit. The correlation between observed and estimated values is 0.96, while the polyno-

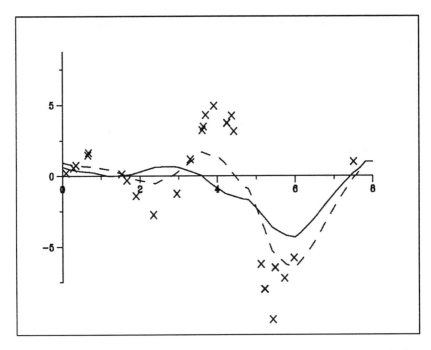

Figure 5.2. The training set x, the response function of the rule system obtained using the counting algorithm (solid line) - and the weighted counting algorithm (dashed line).

mial fit exhibits the somewhat smaller value of 0.89. Another difference between the two fits is that the rule response function is more robust. The results always remain between the minimal and the maximal rule response, in this case in the interval $[-8.21, 9.01]$. This property makes an extrapolation feasible, whereas a polynomial fit cannot be used for extrapolation.

5.4   Partially explicit rule structures

5.4.1   Extension of expert defined rules

An intermediate situation where rule structure is only partially known often occurs — say experts have explicitly formulated knowledge in the form of rules, but the rule system that these experts have constructed is incomplete. In this case, further rules have to be generated using a training set to obtain a complete rule system.

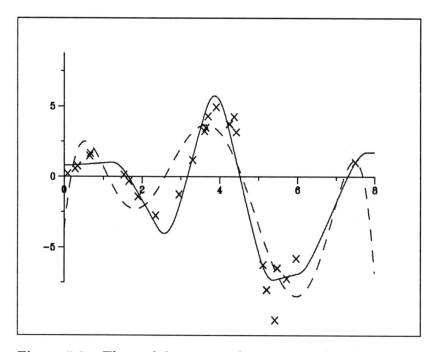

Figure 5.3. The training set x, the response function of the rule system obtained using the least squares algorithm (solid line) and a sixth order least squares polynomial fit to the data (dashed line).

Suppose that the left hand side of the rules can be constructed by the experts so that the rule system becomes complete. The corresponding responses can then be generated using the counting algorithm which is applied only to rules with unknown responses. The same procedure may be applied with the weighted counting algorithm. The least squares algorithm has to be slightly modified.

This property of combining expert knowledge and observed data is an important advantage of fuzzy rule systems. This combination offers an explicit specification of responses in certain areas — and leaves freedom for data "fitting" in other areas. This is essentially different from Bayesian methods – where parameters valid for the entire domain are estimated. This also provides an essential difference from traditional curve fitting techniques, in particular least squares, where constraints such as prescribing function value intervals cannot be considered.

The rule responses can be assessed using the counting and the weighted counting algorithm. There is no change in the method, only rule responses corresponding to the "missing" rules are to be found. The least squares method requires specific modifications.

5.4.2 The modified least squares algorithm

Suppose rules $I_1+1, \ldots, I$ with responses b_i are specified by experts and the responses for rules $1, \ldots, I$ have to be assessed from the training set. The least squares algorithm can be modified to cope with this problem. As in Section 5.3.4 the goal here is to minimize the squared difference between the training set responses and the rule responses:

$$\sum_s \left(\frac{\sum_{i=1}^{I_1} \nu_i(s)M(B_i) + \sum_{i=I_1+1}^{I} \nu_i(s)b_i}{\sum_{i=1}^{I} \nu_i(s)} - b(s) \right)^2 \qquad (5.4.1)$$

Again a necessary and sufficient condition for a minimum of this expression is that its derivative with respect to the unknown $M(B_j)$-s $j = 1, \ldots, I_1$ be 0. Thus for each index j

$$\sum_s \left(\frac{\sum_{i=1}^{I} \nu_i(s)M(B_i) + \sum_{i=I_1+1}^{I} \nu_i(s)b_i}{\sum_{i=1}^{I} \nu_i(s)} - b(s) \right) \frac{\nu_j(s)}{\sum_{i=1}^{I} \nu_i(s)} = 0$$

$$(5.4.2)$$

Rearranging terms, one obtains the linear equation system

$$\sum_s \frac{\sum_{i=1}^{I} \nu_i(s)\nu_j(s)M(B_i)}{(\sum_{i=1}^{I} \nu_i(s))^2} = \sum_s \frac{\nu_j(s)b(s)}{\sum_{i=1}^{I} \nu_i(s)} - \frac{\sum_{i=I_1+1}^{I} \nu_i(s)b_i}{(\sum_{i=1}^{I} \nu_i(s))^2}$$

$$(5.4.3)$$

There are I_1 unknowns $M(B_1), \ldots, M(B_{I_1})$ and I_1 equations, thus the above system usually provides the desired quantities.

Note that ordinary curve fitting methods usually do not allow a subjective prescription of the function values over entire intervals. For example, there is no non-constant polynomial that is constant over an interval. Fuzzy rules provide a tool to analytically model this problem.

5.4.3 Updating a priori given rules

Similarly to the previous situation there are cases where rules can be constructed explicitly by experts, but these should be updated using available data. Another similar case is when rules already assessed should be updated because of possible changes in the system. These cases are different from the previous as there are some rules that are completely known, and some that are not known at all. Suppose prior rule responses $B_1, \ldots, B_I$ are given by experts. The goal is to find updated responses $B_1^*, \ldots, B_I^*$ which do not differ to much from the prior responses, but fit the observed data better. All three previously defined algorithms can be modified for this problem. Only the updating using least squares is described in detail here. As always for the least squares the rule system to be assessed should be used with the normed weighted sum combination method and the mean defuzzification. The first objective is to minimize the squared difference between the rule responses and the observed responses corresponding to the elements of the training set.

$$F_1 = \sum_s \left(\frac{\sum_{i=1}^{I} \nu_i(s) M(B_i^*)}{\sum_{i=1}^{I} \nu_i(s)} - b(s) \right)^2 \qquad (5.4.4)$$

The second objective is that the new responses should not differ much from the prior ones. This can be formulated as:

$$F_2 = \sum_{i=1}^{I} w_i \left(M(B_i) - M(B_i^*) \right)^2 \qquad (5.4.5)$$

Here the weights w_i express the confidence in the prior rules. The weight w_i increases with the confidence in the prior response B_i. The updated responses can be found minimizing the sum of the two objective functions:

$$F_1 + F_2 \rightarrow \min \qquad (5.4.6)$$

This leads to a linear equation system

$$\sum_s \frac{\sum_{i=1}^{I} \nu_i(s) \nu_j(s) M(B_i^*)}{(\sum_{i=1}^{I} \nu_i(s))^2} + w_j M(B_j^*) = \sum_s \frac{\nu_j(s) b(s)}{\sum_{i=1}^{I} \nu_i(s)} + w_j M(B_j)$$

$$(5.4.7)$$

There are I unknowns $M(B_1^*), \ldots, M(B_I^*)$ and I equations, thus the above system usually provides the unknown quantities.

The weights w_i should be assessed in four steps:

1. Express the relative confidence in the rule responses with weights w_i' by the experts without considering the objectives.

2. Calculate the value of the first objective function F_1 using the prior responses B_i. The value of F_1 is denoted as f_1.

3. Solve the linear equation system 5.4.7 using 0 weights ($w_j = 0$ for $j = 1, \ldots, I$). This means that the prior responses are not considered at

all. Calculate the second objective function F_2 using the solution and the weights specified in 1. The corresponding value of F_2 is denoted as f_2.

4. Define the weights w_i using f_1 and f_2 as:

$$w_i = \frac{f_2}{f_1} w_i'$$
(5.4.8)

Other weighting techniques, including subjective ones, can also be applied.

This algorithm not only allows an update of prior rule responses, but also accounts for possible different credibilities.

5.5 Unknown rule structure

It is a frequent problem in modeling to find a mathematical explanation for the behavior of data without any clear prior knowledge. In Bayesian terms this case would lead to the use of a non-informative prior (Berger, 1985). In the case of fuzzy rules the problem is to find the premises among the variables with which a corresponding rule system gives the best explanation of the responses. These types of problems are often treated with multivariate regression techniques (Draper and Smith, 1990). The disadvantage of the regression methods lies in the prescribed form of the response function (usually linear). However, results of a regression analysis can also be used for fuzzy rules — namely, one can choose those variables as premises that were considered for the regression. This way the only problem remaining is the construction of the rule system. There are other possibilities also as described in the next sections.

5.5.1 Elements to be determined

It rather seldom happens that one does not know anything about the system to be modeled. However there may be important explanatory variables that are unknown; furthermore the quantitative correspondence between input and output may also be unknown. Generating rules in such cases is extremely difficult. It consists of two steps: identification of the explanatory variables and then explicit specification of the rules.

Assume that the training set consists of M possible explanatory variables a_m the response b:

$$\mathcal{T} = \{(a_1(s), \ldots, a_M(s), b(s)) \; ; s = 1, \ldots, S\} \qquad (5.5.1)$$

The problem is to find a subset $K \subset \{1, \ldots, M\}$ representing the variables that explain the response, and to construct the corresponding rule system. Here several approaches are available.

The simplest case is when all variables are numeric and measured on a continuous scale. In this situation, the so-called b-cut algorithm can be used.

5.5.2 The b-cut algorithm

The main idea in this algorithm is to partition the response set into distinct classes and find for each hypothesized explanatory variable the ranges corresponding to the partition. If these "inverse images" for a given variable are different for various elements of the partition, then that variable can be used as a premise in the rule system. As usual for fuzzy rule based systems the partition of the response set should include a certain overlap to allow minor discrepancies. The simplest algorithm to accomplish this task is as follows:

1. Find the extreme responses in the training set; so as to define a partition of these responses:

$$b_{\max} = \max_S b(s)$$

and

$$b_{\min} = \min_S b(s)$$

2. Find the extremes for all explanatory variables that are present in the training set:

$$a_{k\,\max} = \max_S a_k(s)$$

and

$$a_{k\,\min} = \min_S a_k(s)$$

These extreme values are used to estimate the explanatory "power" of the variables.

3. Select an integer n. Divide the response range into intervals of equal length $\frac{1}{n}$ over of the entire response range. Since it is necessary that the rules overlap, the intervals are selected so that each point should belong to two intervals. (The exceptions are the endpoints of the range.) Thus, the number of these partitions will be

$$I = 2n - 1$$

The number n has to be selected so that there are enough data points in each partition.

4. Define the support of the possible responses:

$$b_i^- = b_{\min} + \frac{(i - 1)(b_{\max} - b_{\min})}{2n}$$

and

$$b_i^+ = b_i^- + \frac{(b_{\max} - b_{\min})}{n}$$

for $i = 1, \ldots, I$. This step defines the overlapping partition of the response set.

5. Find the variable supports corresponding to the response supports: (Namely find those intervals that deliver values in the selected response partition set.)

$$\alpha_{i,k}^- = \min_{S}\{a_k(s) \text{ such that } b_i^- \le b(s) \le b_i^+\} \tag{5.5.2}$$

and

$$\alpha_{i,k}^+ = \max_{S}\{a_k(s) \text{ such that } b_i^- \le b(s) \le b_i^+\} \tag{5.5.3}$$

The smaller the support (i.e., the difference between $\alpha_{i,k}^+$ and $\alpha_{i,k}^-$) the more the knowledge of a_k helps to determine the corresponding response.

The membership value 1 is assigned to the mean value of the variable corresponding to the response support, that is:

$$\alpha_{i,k}^1 = \frac{1}{N(i)} \sum_{b_i^- \le b(s) \le b_i^+} a_k(s) \tag{5.5.4}$$

where $N(i)$ is the number of such elements s such that

$$b_i^- \leq b(s) \leq b_i^+$$

If $N(i) \leq 1$ or $N(i) << \frac{S}{n}$, then one has to return to Step 3 and repeat the procedure with a smaller value of n.

6. Calculate for each k the relative length of the smallest support:

$$\delta_i = \min_i \frac{\alpha_{i,k}^+ - \alpha_{i,k}^-}{a_{k\,max} - a_{k\,min}} \qquad (5.5.5)$$

7. Variables with

$$\delta_i > \Delta$$

will not be considered as belonging to the rule system; the other ones will constitute the selected set of K explanatory variables. If for a variable the "inverse image" of all partitions is very wide, then it cannot be used as an explanatory variable. The value of Δ should depend on the choice of the number of response classes $(2n - 1)$. Our experience shows that $\Delta \approx \frac{3}{n}$ delivers reasonable rules if $n > 4$.

Once the explanatory variables are selected, one has to construct the left hand side of the rules. Here possible interactions have to be considered. The more interactions are considered the more rules have to be assessed. This step can often be best solved using a trial and error type procedure. For the assessment of the rule responses one can apply one of the algorithms developed earlier for the case of known rule structure and construct the exact rules.

Note that this procedure is in a certain sense the opposite of the counting and weighted counting algorithms. There the left hand sides of the rules are discretized, here the right-hand side. There the image of the classes defines the responses, here the inverse of the responses identifies the variables to be used for the rule system.

5.5.3   Least squares rule estimation

Similarly, as in the case of known rule structure, a least squares rule estimation technique may be used for unknown rule structures. Unfortunately the problem is non-linear and thus in contrast with the known structure case, a simple solution cannot be found. The least squares expression to be minimized is written as:

$$\sum_{s=1}^{S} (R(a_1(s), \ldots, a_K(s)) - b(s))^2 \qquad (5.5.6)$$

This minimum is to be found as a function of the premises $A_{i,k}$ and the responses B_i.

In this case there is no simple linear equation system that yields the solution minimizing (5.5.6). Therefore, other techniques have to be used, depending on the amount and structure of the data.

5.5.4 Other methods

Several authors suggest using a neural network approach for the assessment of rules in the case of unknown rule structures. This is a very promising approach but useful applications were generated mostly in the case of large training data sets that are seldom available in modeling problems. An introduction to neural networks is outside the scope of this book; therefore, the interested reader is invited to refer to other references such as Kosko (1992).

5.6   Deriving rule systems from fuzzy data

The above described methods all assume that the available training set consists of exact (crisp) data. However quite often not only the "knowledge" but the observations are fuzzy as well. The possibility of using rules with fuzzy data was shortly discussed in Chapter 3. If rule systems can use fuzzy premises rule assessment methods using fuzzy data have to be developed. Furthermore, the problem of using fuzzy responses should also be treated. Three cases of fuzzy observations in the training set T can be distinguished:

1. fuzzy premises $\hat{a}_k(s)$

2. fuzzy responses $\hat{b}(s)$

3. fuzzy premises $\hat{a}_k(s)$ and fuzzy responses $\hat{b}(s)$

The treatment of these cases is discussed in the next sections. For the sake of simplicity, only the case of known rule structures is discussed here.

5.6.1 Fuzzy premises

If the premises are fuzzy then one can modify the assessment algorithms. Let $\mu_{\hat{a}_k(s)}(x)$ be the membership function of the fuzzy premise $\hat{a}_k(s)$.

For the *counting algorithm* one has to define a certain threshold level h above which the elements should be considered. Thus the modified algorithm only differs from the original one as follows:

Let R_i be the set of all those alternatives which form a subset of the training set T:

$$R_i = \{(a_1(s), \ldots, a_K(s), b(s)) \; ;$$

$$\max \mu_{\hat{a}_k(s)}(x) > h \; x \in (\alpha^-_{i,k}, \alpha^+_{i,k}) \text{ for all } k = 1, \ldots, K\} \qquad (5.6.1)$$

In this case the calculation of the element with unit membership value requires further efforts. Let $\bar{a}_k(s)$ be the mean of the elements with 1 membership value in the fuzzy set $\hat{a}_k(s)$. Then

$$\alpha_{i,k} = \frac{1}{N_i} \sum_{s \in R_i} \min \left(\max(\bar{a}_k(s), \alpha^-_{i,k}), \alpha^+_{i,k} \right) \qquad (5.6.2)$$

This way the value of $\alpha_{i,k}$ falls into the interval $(\alpha^-_{i,k}, \alpha^+_{i,k})$.

As the sets R_i contain more elements than would be the case of crisp data, the rule responses become fuzzier.

The *weighted counting algorithm* also requires slight modifications. These can be summarized in the following steps.

1. Define the membership functions of the premises as for the counting algorithm.

2. Calculate the DOF-s $\nu_i(s)$ for each element s and each rule i whose premises were defined in Step 1 using Eq. 3.4.1 as:

$$\nu_i(s) = \prod_{k=1}^{K} \max_{x_k} \left(\min(\mu_{\hat{a}_k(s)}(x_k), \mu_{A_k}(x_k)) \right) \qquad (5.6.3)$$

3. Select a number $\varepsilon > 0$. All responses with a DOF of at least ε will be considered in the construction of the rule response. The corresponding response is also assumed to be a triangular fuzzy number $(\beta^-_i, \beta^1_i, \beta^+_i)_T$ where β^+_i is the minimal "answer" with a degree of fulfillment of at least ε

$$\beta_i^- = \min_{\nu_i(s) > \varepsilon} b(s) \qquad (5.6.4)$$

β_i^1 is the DOF weighted mean "answer"

$$\beta_i^1 = \frac{\sum_{\nu_i(s) > \varepsilon} \nu_i(s) b(s)}{\sum_{\nu_i(s) > \varepsilon} \nu_i(s)} \qquad (5.6.5)$$

and β_i^+ is the maximal "answer" with a degree of fulfillment of at least ε

$$\beta_i^+ = \max_{\nu_i(s) > \varepsilon} b(s) \qquad (5.6.6)$$

The value of ε has to be selected so that for each rule a sufficient number of elements of the training set that is considered. As in the case of the counting algorithm, the fuzziness of the responses increases if fuzzy premises are used in the weighted counting algorithm.

The *least squares algorithm* requires very slight modification in this case. The DOF-s of the rules ($\nu_i(s)$) used in the linear equation system 5.3.14 have to be calculated using Eq. 5.6.3. The solution of the equation system then provides the required rule responses. Because of the data with fuzzy premises more rules have positive fulfillment grades and thus the responses are less different than in the case of crisp data.

In large data sets it often happens that there are missing data — observations where not all premises have been measured. These data cannot be well handled in statistical modeling, but by using 1 as the membership function for the missing value they can be treated as data with fuzzy premises.

5.6.2 Fuzzy responses

Fuzzy responses also influence the assessed rule system. The counting and the weighted counting algorithms can be modified as in the previous case. The left-hand side of the rules should be assessed in the same manner as in the crisp case. For the responses in the counting algorithm, one should take:

$$\beta_i^- = \min_{\mu_{b(s)}(b) > 0} b \qquad (5.6.7)$$

$$\beta_i^+ = \max_{\mu_{b(s)}(b)>0} b \qquad (5.6.8)$$

and

$$\beta_i = \frac{1}{N_i} \sum_{s \in R_i} \bar{b}(s) \qquad (5.6.9)$$

where $\bar{b}(s)$ is the mean of the 1 level set of $\hat{b}(s)$. In the case of the weighted counting algorithm, the additional condition of the DOF of the rules should be included, and the same level ε should be applied to the conditions in Equations (5.6.7) and (5.6.8).

For the least squares method the objective is to minimize the squared distance D between the observed and the calculated responses:

$$\sum_s D^2 \left(R(a_1(s), \ldots, a_K(s)) - \hat{b}(s) \right) \qquad (5.6.10)$$

In contrast to the crisp case here the squared distance between a fuzzy and a crisp number has to be calculated. Suppose the responses are triangular fuzzy numbers $\hat{b}(s) = (b_1(s), b_2(s), b_3(s))_T$. In this case the distance measure described in Chapter 2 may be used to minimize the difference between the calculated and the observed responses. Taking a linear weighting function $f(q) = q$ for the distance the function to be minimized becomes:

$$\sum_s (R(a_1(s), \ldots, a_K(s)) - b_2(s))^2 +$$

$$+ \sum_s \frac{1}{3} (R(a_1(s), \ldots, a_K(s)) - b_2(s)) (2b_2(s) - b_1(s) - b_3(s)) +$$

$$+ \frac{1}{12} \left((b_1(s) - b_2(s))^2 + (b_3(s) - b_2(s))^2 \right) \qquad (5.6.11)$$

The response function (as in the crisp least squares case) can be written with the help of the fulfillment grades and the mean rule responses. To minimize the above function with respect to the unknown quantities $M(B_1), \ldots, M(B_I)$ the partial derivatives have to be 0. This leads to the linear equation system:

$$\sum_s \frac{\sum_{i=1}^I \nu_i(s)\nu_j(s)M(B_i)}{(\sum_{i=1}^I \nu_i(s))^2} =$$

$$= \sum_s \frac{\nu_j(s)}{\sum_{i=1}^I \nu_i(s)} \left(\frac{1}{6} b_1(s) + \frac{2}{3} b_2(s) + \frac{1}{6} b_3(s) \right) \qquad (5.6.12)$$

Note that the left hand side of the equation system is the same as in the crisp case (Equation 5.3.14). On the right hand side the crisp answer $b(s)$ is replaced with the crisp number which is the closest to the fuzzy response with respect to the selected distance D.

5.6.3 Fuzzy premises and fuzzy responses

The most frequent case in assessing rules from fuzzy data is the case where both the premises and the responses are fuzzy. For this purpose the counting and the weighted counting algorithms can be modified according to the two previous sections. The left hand side of the rules should be assessed as in the case of fuzzy premises, the right hand side as in the case of fuzzy responses. For the least squares algorithm the equation system depends on the distance measure used. The linear equation system remains the same as for the fuzzy response case (5.6.12), but the DOF-s $(\nu_i(s))$ have to be calculated according to the case of fuzzy premises (Equation 5.6.3).

5.6.4 Additional remarks

Extension of rules and update of rule systems with the help of fuzzy data can be done in a similar manner as in the crisp case. The modification of the algorithms can be done using the same ideas as presented in the previous sections.

A mix of fuzzy and crisp data causes no problems in applying the above methods.

5.7 Rule verification

Explicit rules contain expert knowledge and thus can be assumed to be correct. Even then and much more so if a rule system is derived from a training set, it has to be verified. For this verification a split-sampling approach is suggested as in Blinowska et al. (1992). Let

$$P = \{(a_1(s), \ldots, a_K(s), b(s)) \ ; s = 1, \ldots, S\} \qquad (5.7.1)$$

be the set of all available observations. This set is divided into two disjoint subsets, the training set $\mathcal{T}$ and the verification set $\mathcal{V}$:

$$P = \mathcal{V} \cup \mathcal{T}$$

$$\mathcal{V} \cap \mathcal{T} = \emptyset$$

Rules are derived from $\mathcal{T}$ and then responses $R(a_1(s), \ldots, a_K(s))$ are calculated for all elements s of $\mathcal{V}$. The quality of the rule system may be measured by the mean error:

$$E = \frac{1}{|\mathcal{V}|} \sum_{s \in \mathcal{V}} D\left[R(a_1(s), \ldots, a_K(s)), b(s)\right] \qquad (5.7.2)$$

where D is a measure of distance between calculated response R and observed response b. In the case of numerical rules traditional measures can be used for the verification. These are the mean error, which is a measure of bias

$$E_1 = \frac{1}{|\mathcal{V}|} \sum_{s \in V} \left[R(a_1(s), \ldots, a_K(s)) - b(s)\right] \qquad (5.7.3)$$

the maximum absolute deviation, which is an upper limit of the error:

$$E_2 = \max |R(a_1(s), \ldots, a_K(s)) - b(s)| \qquad (5.7.4)$$

the sum of absolute deviations, which is one measure of accuracy:

$$E_3 = \frac{1}{|\mathcal{V}|} \sum_{s \in V} |R(a_1(s), \ldots, a_K(s)) - b(s)| \qquad (5.7.5)$$

and the sum of squared errors, which is another measure of accuracy, more sensitive to large deviations than the absolute value sum:

$$E_4 = \frac{1}{|\mathcal{V}|} \sum_{s \in V} \left[R(a_1(s), \ldots, a_K(s)) - b(s)\right]^2 \qquad (5.7.6)$$

A further measure is the correlation coefficient of the observed and calculated rule responses which is calculated as:

$$\rho = \frac{\dfrac{\sum R(a_1(s),...,a_K(s))b(s)}{|\mathcal{V}|} - \dfrac{\sum R(a_1(s),...,a_K(s))\sum b(s)}{|\mathcal{V}|^2}}{\sqrt{\dfrac{\sum R(a_1(s),...,a_k(s))^2}{|\mathcal{V}|} - \left(\dfrac{\sum R(a_1(s),...,a_k(s))}{|\mathcal{V}|}\right)^2}\sqrt{\dfrac{\sum b(s)^2}{|\mathcal{V}|} - \left(\dfrac{\sum b(s)}{|\mathcal{V}|}\right)^2}}$$

(5.7.7)

The summations in the above equation take place over all the elements of set $\mathcal{V}$. The values of these performances indices provide prior measures of performance of the rule system applied to an arbitrary sets of premises.

Example 5.2 *Consider the case of Example 5.1. The training set given in Table 5.4 was obtained after a random split sampling of the entire data set containing 50 values. The remaining 25 data points are listed in Table 5.8. Rule systems assessed with the counting, weighted counting*

Table 5.8.   Verification set $\mathcal{V}$

s	$a(s)$	$b(s)$	s	$a(s)$	$b(s)$
1	0.04	0.08	14	4.07	5.05
2	0.27	0.42	15	4.14	4.92
3	0.44	1.04	16	4.63	0.94
4	0.56	0.82	17	5.32	-7.07
5	0.79	1.32	18	5.33	-6.34
6	0.90	1.66	19	5.51	-10.44
7	1.10	1.19	20	5.69	-7.16
8	1.17	1.19	21	5.87	-6.72
9	1.61	-0.17	22	5.98	-7.70
10	3.02	-0.79	23	6.10	-4.60
11	3.11	-0.25	24	7.43	0.71
12	3.50	2.59	25	7.88	1.68
13	3.94	4.53			

and least squares algorithm were validated using the verification (or validation or testing) set $\mathcal{V}$ (Table 5.8). For comparison purposes the same performance indices were calculated for the training set $\mathcal{T}$ (Table 5.4). Table 5.9 shows the different performance indices for the training set and the verification set for the rule systems generated by the counting, the weighted counting and the least squares algorithms. For a further comparison, the rule combination methods were varied for the counting and the weighted counting algorithms. Such a sensitivity analysis cannot

*be performed for the least squares algorithm as it only delivers the fuzzy
means of the response. Mean defuzzification was used in conjunction with
each combination method. In the case of the maximum and the weighted
sum combination method the crisp answer of Rule 7 could be neglected.
To avoid this occurrence, the answer was changed to $(0.9, 1, 1.1)_T$. This
change had no effect on the results already presented in Example 5.1.
All rule systems considered were successfully verified. There is only a*

Table 5.9. **Performance indices for the different rule assessment techniques and combination methods (all with mean defuzzification)**

Method	Data set	E_1	E_2	E_3	E_4	ρ
Counting alg.	T	0.28	6.49	2.67	10.99	0.71
Normed w-sum.	V	0.16	6.63	2.22	8.68	0.81
Counting alg.	T	0.50	6.71	3.00	13.10	0.56
W-sum.	V	0.41	6.85	2.52	10.54	0.70
Counting alg.	T	0.47	7.10	3.19	14.28	0.48
Maximum	V	0.43	7.08	2.60	10.84	0.68
W.-counting alg.	T	0.22	4.84	1.42	3.51	0.93
Normed w-sum.	V	0.13	4.81	1.74	5.00	0.90
W.-counting alg.	T	0.52	7.55	2.53	11.07	0.64
W-sum.	V	0.46	7.50	2.23	9.40	0.71
W.-counting alg.	T	0.40	6.75	2.32	9.80	0.69
Maximum	V	0.44	6.66	2.59	11.41	0.63
Least squares	T	0.13	2.97	0.90	1.62	0.96
Normed w-sum.	V	0.25	3.15	0.70	1.08	0.97

*small difference between their performance when using the training set
T and the verification set V. In several cases the verification set happened to be better modeled by the rule system than the training set. The
least squares algorithm delivered the best performance indices. The rule
system obtained using the weighted counting algorithm performed better
than the one obtained from the counting algorithm. The normed weighted
sum combination was better than the weighted sum or the maximum
combination. The sixth order polynomial fitted to the training set T in
Example 5.1 was also verified. Table 5.10 shows the corresponding performance indices. One can observe that the polynomial delivers a good
fit of the training set, but its performance is poorer in contrast on the*

verification set. The least squares rule estimation was superior in both training and verification phases of the analysis.

Table 5.10. **Performance indices for least squares polynomial fitting**

Method	Data set	E_1	E_2	E_3	E_4	ρ
Sixth-order	$\mathcal{T}$	*0.00*	*3.58*	*1.54*	*3.27*	*0.89*
polynomial fit	$\mathcal{V}$	*0.42*	*5.05*	*1.72*	*4.28*	*0.87*

The results of this example once again support our recommending the use of normed weighted sum combination and mean defuzzification for constructing fuzzy rule systems.

5.8   Removing unnecessary rules

As already pointed out, an essential advantageous property of fuzzy rule systems is the overlap in the definition of the premises. Even though this is a very desirable property for operational purposes it is useful to remove rules that only have a small influence on the final response function. Such a removal simplifies the model and thus improves its computational performance. This is very important in fuzzy control.

A rule can only be removed if the rule system remains complete even without this rule. As pointed out in Proposition 4.5, if:

$$O_v(a_1, \ldots, a_K) \geq 2 \quad \text{for all} \quad (a_1, \ldots, a_K) \in \text{supp}(i_0) \cap \mathcal{A}$$

then the rule system remains complete after the removal of rule i_0. The removal of a rule from the rule system increases the fuzziness of the rule response for the minimum combination methods. In contrast, for the maximum and the additive combinations, the rule response fuzziness decreases.

Rule verification measures can be used to decide whether a rule has an important role or not.

For the preferred additive combinations and fuzzy mean defuzzification procedures, the effect of removing a rule can be calculated directly. Consider the the normed weighted sum combination method and the mean defuzzification. Let $F(a_1, \ldots, a_K)$ be the sum of the DOF-s of all rules applied to the premise vector $(a_1, \ldots, a_K)$. Formally:

$$F(a_1, \ldots, a_K) = \sum_{i=1}^{I} D_i(a_1, \ldots, a_K) \tag{5.8.1}$$

The response function for the normed weighted sum combination and the mean defuzzification is:

$$R(a_1, \ldots, a_K) = \frac{\sum_{i=1}^{I} D_i(a_1, \ldots, a_K) M(B_i)}{F(a_1, \ldots, a_K)} \tag{5.8.2}$$

Removing rule i_0 the response function is:

$$R^{(i_0)}(a_1, \ldots, a_K) = \frac{\sum_{i=1}^{I} D_i(a_1, \ldots, a_K) M(B_i) - D_{i_0}(a_1, \ldots, a_K) M(B_{i_0})}{F(a_1, \ldots, a_K) - D_{i_0}(a_1, \ldots, a_K)} \tag{5.8.3}$$

Thus the effect of rule i_0 is the difference between the two response functions:

$$|R(a_1, \ldots, a_K) - R^{(i_0)}(a_1, \ldots, a_K)| =$$

$$\left| \frac{D_{i_0}(a_1, \ldots, a_K) \left(\sum_{i=1}^{I} D_i(a_1, \ldots, a_K) M(B_i) - M(B_{i_0}) F(a_1, \ldots, a_K) \right)}{F(a_1, \ldots, a_K) \left(F(a_1, \ldots, a_K) - D_{i_0}(a_1, \ldots, a_K) \right)} \right| \tag{5.8.4}$$

Note that this difference is 0 outside the support of rule i_0. Thus the difference has to be calculated only on the support of the rule that is considered for removal.

If the maximum of this difference is small enough, then the rule i_0 can be removed from the rule system. The calculations in the weighted sum combination and mean defuzzification cases are similar.

Note that if the least squares algorithm is used for rule assessment, then a rule assessment without rule i_0 can yield a smaller difference in the response function than calculated above. This is not the case for the counting and the weighted counting algorithm.

Example 5.3 *Consider the rule system of Example 5.1 obtained using the least squares algorithm. Table 5.7 contains the rules. For example, Rule 2 could be removed, as its support is fully contained in the supports of Rules 1 and 3. To see the effect of the removal the difference between the two rule response functions has to be calculated on the support of*

Rule 2 which is the interval $(0,3)$. *For this purpose the DOF functions of the four rules are calculated:*

$$D_1(a) = \begin{cases} \frac{a+1.2}{1.95} & 0 \le a < 0.75 \\ \frac{1.8-a}{1.05} & 0.75 \le a \le 1.8 \end{cases}$$

$$D_2(a) = \begin{cases} \frac{a}{1.24} & 0 \le a < 1.24 \\ \frac{3-a}{1.76} & 1.24 \le a \le 3 \end{cases}$$

$$D_3(a) = \begin{cases} \frac{a-1.2}{1.66} & 1.2 \le a < 2.86 \\ \frac{4.2-a}{1.34} & 2.86 \le a \le 3 \end{cases}$$

$$D_4(a) = \frac{a-2.4}{1.64} \quad 2.4 \le a \le 3$$

The rule response function using all three rules is

$$R(a) = \frac{0.8D_1(a) + 0.95D_2(a) - 5D_3(a) + 9.01D_4(a)}{D_1(a) + D_2(a) + D_3(a) + D_4(a)}$$

Using Rules 1 and 3 only:

$$R^{(2)}(a) = \frac{0.8D_1(a) - 5D_3(a) + 9.01D_4(a)}{D_1(a) + D_3(a) + D_4(a)}$$

The difference between the two responses can be calculated using (5.8.4). Simple calculations give that this maximum is attained at $a = 1.8$ *and the maximal difference is 3.89. Thus, the removal of the rule influences the response function considerably. If one removes Rule 1, then the rule response function becomes:*

$$R^{(1)}(a) = \frac{0.95D_2(a) - 5D_3(a) + 9.01D_4(a)}{D_2(a) + D_3(a) + D_4(a)}$$

In this case the maximal difference on the interval $(0,3)$ *is only 0.2 attained at* $a = 1.55$. *Thus, the removal of Rule 1 does not considerably change the response function. Figure 5.4 shows the three rule response functions, supporting the possibility of the removal of Rule 1.*

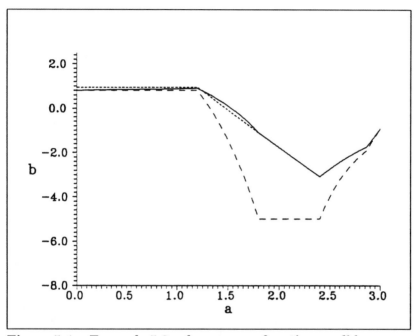

Figure 5.4. Example 5.3 rule response functions: solid = complete rule system; dashed (long dashes) = without rule 2; dashed (short dashes) = without rule 1

6

Fuzzy rule-based modeling versus fuzzy control

In this chapter a brief reminder of the principles of elementary fuzzy control are given and then related to the type of fuzzy rule-based modeling that has been introduced to this point. Fuzzy control has proven to provide simple and stable systems and is being applied to many practical and industrial uses. We began our research by using the ideas of fuzzy control and had to design better procedures because fuzzy rule-based modeling has to be done in a quasi open loop mode. This is because trial and error adjustment of procedures (rules descriptors), which is akin to system identification, cannot be done in real time, as explained in this chapter.

Control theory is the formalization and generalization of actions that individuals take every day. Consider for example how we seek to regulate the temperature of a shower by means of a mixer providing constant flow. The input is the setting of the mixer lever and the output is the temperature of the water. This simple system is only operated in a "closed loop" manner, with fuzzy rules of the type "If the water is (much colder, somewhat colder, much hotter than desired, turn the lever to the right or the left to a certain extent". Should the relative temperatures of the hot and cold water change, the observation of the output temperature is compared to the desired temperature and the difference is used as a feedback to set the mixing lever. In a similar fashion, a thermostat compares the setting (a desired output) with the actual temperature: it turns the heating on if the difference exceeds a certain threshold and turns it off if it is within a certain tolerance. Either problem can be modeled by a fuzzy rule-based approach.

Traditionally automatic control is based on engineering control theory. The control system consists of a control algorithm which uses the process input and the observations of the previous state of the process (feedback) to find an optimal control applied to the system in the next time step. Measurement inaccuracies or incomplete observability of the process state make the control of complex processes extremely difficult. Figure 6.1 shows the structure of a traditional controller.

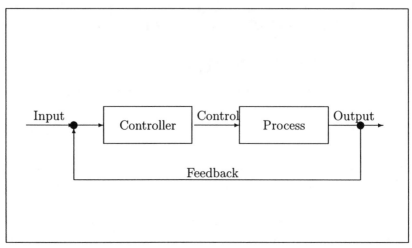

Figure 6.1.   The traditional controller.

It is rather paradoxical that fuzzy sets that express imprecision and uncertainty are so frequently used to determine the very precise actions required by control. However, it seems that fuzzy rules provide a useful tool to obtain fast control for systems with continuous sensor based observations. The speed of the control, gained through the simplicity of the rules, fully compensates for the minor inaccuracies of the system.

Traditional control theory is being applied successfully in a great number of cases. For complex problems, such as the driving of a car, control theory cannot be successfully applied. However, an average person after some training can handle a car — even without any fundamental understanding of the underlying physical principles. This shows that even to control a complex system it is often enough to have qualitative knowledge about its behavior. Complex industrial processes are often better controlled by an experienced human controller than by automatic control. This property is successfully used in fuzzy control, where instead of the exact equations fuzzy rules are used.

6.1   Principles of fuzzy control

Fuzzy control differs from traditional control in its very foundation. In traditional control schemes, the process was first modeled with the help of physical or chemical relationships. This model was subsequently used to design the control strategy.

Originally fuzzy control had a less ambitious approach in modeling the system — it simply strived to mimic the functioning of a human operator. In subsequent development, researchers such as Sugeno (1985) have attempted to construct fuzzy controllers on the basis of fuzzy models of systems, and are even using training sets to calibrate their controller.

Returning to the original type of fuzzy control that inspired our research, there was no attempt to arrive at a complete understanding of the process it tries to react to the input and the feedback using rules based on knowledge and experience. The construction of a fuzzy controller consists of three steps. First, the input is transformed into rule arguments that may be crisp or fuzzy depending on the imprecision and accuracy of the control to be obtained. In the second step, fuzzy rules are applied to transform the arguments into a fuzzy response. In this step the DOF-s are calculated and combined according to a selected inference and combination method (Chapter 3) yielding a fuzzy control scheme. The third and final step consists in the defuzzification of the fuzzy control to obtain a crisp executable control action (Sugeno 1985). This step is performed by applying an appropriate defuzzification method (see Chapter 3). Figure 6.2 shows the elements of fuzzy control.

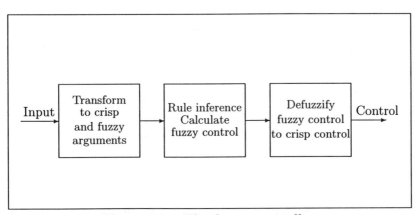

Figure 6.2.   The fuzzy controller.

The fuzzy controller seems to be more complicated than a traditional controller, for which neither the first transformation nor the last defuzzification are necessary. In fact the fuzzy controller follows the same steps as a human controller — it reads a numeric value, transforms it into a qualitative description (high pressure), finds the appropriate qualitative action(s) for this qualitative case, and finally transforms the qualitative action into a real action and executes it. It is possibly this aspect that has made fuzzy control extremely successful in the last few years. Another great advantage of fuzzy control in real-time complex situations is its high speed of response because of simple mathematical calculations, in contrast with algebraic control models. For example, the automatic control of a portable video recorder (camcorder) would hardly be possible using standard control theory. With fuzzy control, commercial applications have become possible. In the next section, a few examples illustrate the functioning of a fuzzy controller.

6.2   Examples of fuzzy control

There are a great number of successful applications of fuzzy control (Sugeno 1985). As an illustration, a few examples taken from the literature are described. The first example uses the inverted pendulum to illustrate the principles of fuzzy control. In the next example, automatic train control is briefly described.

6.2.1   Inverted pendulum

The example of the inverted pendulum given in Yamakawa (1989) and again in Rommelfanger (1993) is selected to illustrate elementary fuzzy control principles. Consider the problem of keeping an inverted pendulum (which is fixed) articulated at a fixed point on a mobile cart. The cart can move forward and backward, and the controller decides on the direction and acceleration of the cart.

This inverted pendulum makes an angle θ with the vertical direction and is to be kept vertical by a lateral motion, assuming, for simplicity, that the system possesses only one degree of freedom:

$$\dot{x} = \frac{dx}{dt}$$

As a function of angle θ and angular velocity one finds:

$$\dot{\theta} = \frac{d\theta}{dt}$$

A sensor then measures $\theta, \dot{\theta}$ and $\dot{x}$ has to be adjusted via a real time feedback loop. While the classical equations of motion of such a system are extremely complicated and depend upon the specific characteristics of the pendulum (mass distribution, length), the reference cited shows that the fuzzy rules given below provide a stable fuzzy control of the pendulum independently of its characteristics. The idea of fuzzy control

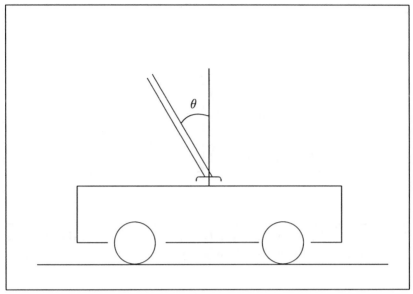

Figure 6.3.   The cart with the inverted pendulum

in this case is that the cart has to be moved into the opposite direction of the pendulum; the opposing motion should increase as a function of increasing angle θ and angular velocity $\dot{\theta}$. If θ and $\dot{\theta}$ are of opposite sign, the motion should be moderate to zero. This simple principle does not require any detailed physical description of the problem.

To define the rule let the possible states of $\dot{x}, \theta, \dot{\theta}$ be:

positive medium (PM)
positive small (PS)
about zero (ZR)
negative small (NS)
negative medium (NM)

The "positive" reflects a selected direction, "negative" is the opposite direction. Two other states may be defined, positive large (PL) and negative large (NL) but Yamakawa (1989) demonstrates that they are either not needed or not useful. To each of these linguistic descriptors one associates a triangular fuzzy number with membership function as shown in Figure 6.4. One could as well use trapezoidal membership functions instead of triangular ones.

For each combination of the states (or premises) $(\theta, \dot{\theta})$, a corresponding action (or consequence) $\dot{x}$ can be selected, leading to a rule system

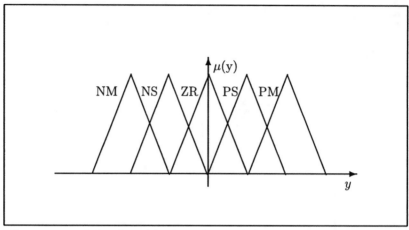

Figure 6.4.   Membership functions of the linguistic descriptors where y stands for $\theta, \dot{\theta}$ or $\dot{x}$, with proper units.

with up to 5×5=25 rules. In fact it can be demonstrated that not all rules are needed. Using the descriptors given before, it is found that seven rules numbered 1 to 7 in Table 6.1 are necessary and sufficient for stable control.

Table 6.1.   Rules for stable control of the inverted pendulum

Rule n°	If θ is	and $\dot{\theta}$ is	then select $\dot{x}$ as
1	ZR	ZR	ZR
2	NS	PS	ZR
3	NS	NS	NS
4	NM	ZR	NM
5	PS	NS	ZR
6	PS	PS	PS
7	PM	ZR	PM

The rule system is applied in the same way as described before in Chapter 3: given a sensor measured state $(\theta, \dot{\theta})$, the DOF of each rule is calculated using the product or min-max inference. Rule responses are then combined — usually either the maximum or the cresting maximum combinations are used in fuzzy control. Finally the combined response is defuzzified using the fuzzy mean to obtain the crisp control value of $\dot{x}$.

The membership functions are given subjectively and then adjusted by trial and error via the feedback control. Notice that the rule system can be fine-tuned by further experiments, which is seldom the case in fuzzy rule based modeling.

6.2.2   Automatic train operation system

This example is a real-life application of fuzzy control to the control of the subway system of the city Sendai (Japan). The great number of passengers traveling on the subway requires a safe, efficient and comfortable control system. Fuzzy control seems to be appropriate for this purpose, as the continuous trajectories it generates smooth braking and acceleration. The description of the system is based on Yasunobu and Miyamoto (1985).

The fuzzy control is designed to obtain a control strategy similar to that of an experienced human operator. It consists of two major functions:

1. the constant speed control (CSC) to keep the speed below a specified upper limit.

2. the train automatic stop control (TASC) to regulate the speed in order to stop at a prescribed location.

For both functions a set of rules are developed based on the experience of the human controller. A control strategy is evaluated with the help of fuzzy performance indices that form six groups:

1. Safety: first and foremost, design a safe system.

2. Comfort: the passengers should feel comfortable without sudden acceleration, deceleration, or jerks.

3. Traceability: train should adhere to set schedule and speed ranges.

4. Energy saving: smooth operation should save energy.

5. Running time: the train should run as fast as possible.

6. Stop gap: the train should stop as close as possible to a prescribed point.

These performance indices are evaluated continuously during the operation. So called predictive fuzzy control rules are developed which describe the control action (a change of the powering or on the braking notch) using the present control action and the different performance indices to define the new control action. Separate rules are developed for the CSC and the TASC. An exact definition of the performance indices and a set of operating rules are listed in Yasunobu and Miyamoto (1985). The fuzzy control operates robustly and realizes a very smooth control with a small number of notch changes. Furthermore it saves over 10% of the energy consumed by conventional controllers.

6.3   Fuzzy control and fuzzy rule-based modeling

Elements for this comparison, which have essentially been presented throughout the preceding material, are now summarized.

6.3.1   Closed loop versus open loop

Fuzzy control is a closed loop procedure in real-time allowing observation at each instant of the true value of the state variable vector to determine the new value of the control. This feedback can correct minor errors in the rule system. In contrast, fuzzy rule-based modeling is a quasi open loop procedure which may in fact be considered as system identification: here only the complete trajectory of the state or output variable may be observed, so that small differences may lead to widely divergent trajectories; yet in every case we have studied, our fuzzy rule-based modeling approach appears to be extremely robust in terms of the difference between predicted and observed values staying small.

6.3.2   Fuzzy set membership function assessment

In most fuzzy control cases, fuzzy set membership functions corresponding to the various possible state descriptions are determined subjectively and then adjusted by trial and error or else by artificial neural nets (Tagaki, 1990). In each of these cases, the feedback-loop is used to determine the adjustment. In fuzzy rule-based modeling, the fuzzy set membership function may be assessed subjectively, from a physical model, from data, or from a combination thereof (Blinowska and Duckstein, 1993). Neural networks may also be used here (Tagaki and Hayashi, 1990), but trial and error cannot be used on a real time basis.

6.3.3   Derivation of rules

The rules in fuzzy control can be derived from:

1. the operator's experience
2. the operator's control actions

3. a crisp or fuzzy model of the process to be controlled

4. training sets

A most common approach appears to be the first one, using the subjective input of control specialists, such as train operators. The second approach is used in industrial problems. As an example of the third approach, Sugeno (1985) uses a fuzzy model of the system to be controlled, to construct the controller. Finally, (Mamdani et al., 1983) use training set to design fuzzy controllers.

In fuzzy rule-based modeling, it is also possible to derive rules subjectively from accepted models or explicit rules; however, a training set can be used to establish a set of implicit rules derived either by a subjective procedure or from a physical model. The rule assessment techniques described in Chapter 5 can also be used for fuzzy control. There are several learning algorithms developed for the assessment of rule systems for designing fuzzy controllers. The Q-learning method is an example where the different controllers are evaluated using quality criterion functions.

Speed is an important factor for controllers; thus, the reduction of the rule system to a minimum is often required. This can be done using trial and error. Alternatively, methods described in Ssection 5.8 can also be used for this purpose. Furthermore, even if the speed does not play a central role in the control, a reduced rule system can often be realized at a lower cost, which is important for devices produced in great quantities. In rule-based modeling, the size of the rule system does not play a central role, but a parsimony modeling principle should still be used. The accuracy of the rules is a more important criterion than execution time because of the open loop nature of the trajectories which might diverge in case of "unbalanced" rules.

In fuzzy rule-based modeling the available observations are usually limited, and the rules have to be assessed without any further data. In fuzzy control, the control actions can be tested without important limitation. Specific observations can be performed to test the control. This nearly "unlimited" availability of data makes it possible to combine neural net and fuzzy techniques in control, which is usually not the case in fuzzy rule-based modeling.

A further difference in rule assessment is the adaptive learning capability of fuzzy control. This means that by observing the performance of the controller, the control rules can be modified. This is usually not possible in fuzzy rule-based modeling.

6.3.4   Rule combination and defuzzification

In fuzzy control, usually the maximum or the cresting maximum combination of rule responses is used, and the defuzzification procedure commonly applied is by fuzzy mean (center of gravity). Since the rule systems in fuzzy control are reduced as much as possible to their minimum form, the choice of combination and defuzzification methods do not play a central role. The simplicity of realization or availability of the necessary hardware plays a more important role.

We do not pretend to provide even a brief overview of the various facets of fuzzy control, since literature on this subject is most abundant; we merely hope to have pointed out how and why our fuzzy rule based modeling differs from a standard elementary fuzzy control, even if there are many apparently similar features. In fact, we suggest that parts of our procedure could advantageously be used in fuzzy control: for example our recommended defuzzification scheme would present in any case the advantage of simplicity of calculation, since center of gravities of each membership functions have to be calculated once and for all, and calculating the area of a triangle or of a trapeze is most simple. Recently, Wang and Chen (1993) have shown that the defuzzification method makes little difference, if any, in the performance of certain adaptive control systems, so that one may as well use the simplest technique; his findings thus agree with the results that we have partly presented in Chapter 4.

6.3.5   Further remarks

In fuzzy control much emphasis has been given to the development of hardware that realizes rule-based systems (Zhang et al. 1993). Real time control and/or mass production requires such a view point. In contrast, modeling is done on a more individual basis so that hardware-based performance improvements are not usually needed. Yet even in fuzzy rule-based modeling, it is important to avoid an excessive redundancy of rules, which may introduce noise into the model and generate "flat" responses.

As a final remark, we wish to emphasize that if fuzzy rule-based modeling is usually an open loop procedure, it is not often akin to an open loop control model. The procedure must first identify the system and then reproduce its operation in a validation phase. An exception to this may be reservoir operation, which is in fact an open loop control model, as shown in Chapter 10. The control state of reservoir is the water volume stored and the variable, water released (or opening of the gate). Feedback is delayed until at least the next time step, which may be one week. The premises in that model are water inflow and demand prediction, as well as system state (storage).

7

Rule systems with discrete responses

In Chapter 4 properties of numerical rule systems were investigated in detail. In quite a few cases rule systems are constructed with responses defined over elements of a discrete set (usually finite, sometimes even binary). If a crisp final answer is to be found, the procedure is really analogous to a classification problem since a specific consequence element has to be assigned to every premise vector $(a_1, \ldots, a_K)$. As pointed out in Chapter 3 one should use the maximum defuzzification method in these cases. A mean or median defuzzification could only be used if the discrete set is numeric and even in this case they could lead to responses that do not belong to the discrete response set. An intermediate value may not have any meaning. The classification with fuzzy rules is different from the classifications obtained with fuzzy pattern recognition (Bezdek 1981). In pattern recognition, the classes (responses) also have to be identified in finding relationships between the data. Fuzzy rules are to be used if the groups that form the elements of the finite set of responses are already defined and an explanation of the classification of other objects is to be done.

Furthermore it should be kept in mind that Proposition 4.4 and Corollary 4.1 do not hold into the discrete response case; thus the rules have to be formulated with the logical "OR", "AND", "XOR", "MOST OF" and "AT LEAST A FEW" operators as defined in Chapter 3.

7.1 Combination of discrete consequence type rules

Let the discrete set $\{b_1, \ldots, b_n, \ldots\}$ be the common set in which all the rule responses are defined as fuzzy subsets. The consequence or response set corresponding to rule i is described as a two-tuple description of the consequence set element:

$$B_i = ((b_1, \mu_i(1)), (b_2, \mu_i(2)), \ldots)$$

where $\mu_i(n)$ is the membership of b_n in the fuzzy set B_i. Let $\nu_i = D_i(a_1, \ldots, a_K)$ be the degree of fulfillment of rule i. The problem is to find the rule response set B, which depends on the combination method:

$$B = ((b_1, \mu(1)), (b_2, \mu(2)), \ldots)$$

Using the maximum defuzzification method, only the element $(b_{\max}, \mu(\max))$ of B has to be identified for which:

$$\mu(\max) \geq \mu(i) \quad \text{for all} \quad i \tag{7.1.1}$$

In each combination case this element may be found in a different manner.

7.1.1 Minimum combinations

The minimum combination rule system yields the response set:

$$B = \left((b_1, \min_{\nu_i > 0}[\nu_i \mu_i(1)]), (b_2, \min_{\nu_i > 0}[\nu_i \mu_i(2)]), \ldots \right) \tag{7.1.2}$$

where the minimum is taken over all indices i for which the rule i has a positive DOF ($\nu_i > 0$). The maximum defuzzification finally selects the element with the highest membership value. The verbal description of this method is: The element that contradicts all the applicable rules to the lowest degree is selected as representative.

7.1.2 Maximum combinations

The maximum combination yields the response set:

$$B = \left((b_1, \max_{i}[\nu_i \mu_i(1)]), (b_2, \max_{i}[\nu_i \mu_i(2)]), \ldots \right) \tag{7.1.3}$$

where the maximum is taken over all indices i.

As this is a classification problem, each rule should have a single most preferred consequence. Let $b_i(m_i)$ be the element for which

$$\mu_i(m_i) > \mu_i(n) \quad \text{if} \quad n \neq m_i$$

This $b_i(m_i)$ is the most preferred answer of rule i. In this case the maximum combination method gives the defuzzified answer:

$$b(\max) = b(m_s) \quad \text{if} \quad \nu_s\mu_s(m_s) > \nu_i\mu_i(m_i) \quad \text{for} \quad i \neq s \qquad (7.1.4)$$

Note that in the case of this combination method, only the most preferred element is used in each rule to find the defuzzified answer. The response of each rule can thus be defined as a single element, as no element having a smaller membership value is considered in Equation (7.1.4). Consequently the response set B_i can be replaced by

$$B_i' = (\dots (b_{m_i-1}, 0), (b_{m_i}, \mu_i(m_i)), (b_{m_i+1}, 0) \dots)$$

without changing the final consequence. It is reasonable to formulate the rule system with crisp answers, usually with a membership value $\mu_i(m_i) = 1$. This means that a fuzzy rule system with a discrete response set, maximum combination, and maximum defuzzification can be formulated as a fuzzy rule system with crisp responses. The rules in this case describe the classes, and are thus easier to be formulated by experts. The response corresponding to the rule with the highest DOF is assigned as response by the rule system. This kind of rule system seems to be most suitable for discrete response cases.

7.1.3 Additive combinations

The weighted sum combination of the rule system yields to the response set:

$$B = \left((b_1, \frac{\sum_i[\nu_i\mu_i(1)]}{M_1})(b_2, \frac{\sum_i[\nu_i\mu_i(2)]}{M_1}), \dots \right) \qquad (7.1.5)$$

where:

$$M_1 = \max_n \sum_i [\nu_i\mu_i(n)]$$

The normed weighted sum combination of the rule system yields to the response set:

$$B = \left((b_1, \frac{\sum_i [\nu_i \beta_i \mu_i(1)]}{M_2})(b_2, \frac{\sum_i [\nu_i \beta_i \mu_i(2)]}{M_2}), \ldots \right) \tag{7.1.6}$$

where β_i is the inverse of the cardinality of B_i as defined in Chapter 2 definition 2.2:

$$\beta_i = \frac{1}{\mathrm{car}(B_i)} = \frac{1}{\sum_j \mu_i(j)}$$

and

$$M_2 = \max_n \sum_i [\nu_i \beta_i \mu_i(n)]$$

In these cases even an element which was not most preferred in any of the individual rules may turn up as a best "compromise". The additive combination can also be used in the case of rules with crisp responses. If for each element in B there is a single rule with that response, the additive and the maximum combinations yield the same result after maximum defuzzification.

Example 7.1 *Consider the rules*

$$If \ (1,2,3)_T \ AND \ (1,2,6)_T \ then \ ((b_1, 0.1), (b_2, 0.7), (b_3, 1.0))$$

and

$$If \ (2,3,4)_T \ AND \ (2,4,6)_T \ then \ ((b_1, 1.0), (b_2, 0.8), (b_3, 0.1))$$

Taking $a_1 = 2.5$ and $a_2 = \frac{10}{3}$ the DOFs of the first and second rule are the same and equal to $\nu_1 = \nu_2 = \frac{1}{2}.\frac{2}{3} = \frac{1}{3}$. The maximum combination method yields $((b_1, \frac{1}{3}), (b_2, \frac{4}{15}), (b_3, \frac{1}{3}))$. Thus, the defuzzification is not unique — b_1 and b_3 have the same membership value in the combined set.

The minimum combination leads to $((b_1, \frac{1}{30}), (b_2, \frac{7}{30}), (b_3, \frac{1}{30}))$. The defuzzification leads to b_2.

The weighted sum combination yields using Eq. 7.1.5 the response set B:

$$B = \left[(b_1, \frac{\mu_1(1) + \mu_2(1)}{2M_1}, (b_2, \frac{\mu_1(2) + \mu_2(2)}{2M_1}, (b_3, \frac{\mu_1(3) + \mu_2(3)}{2M_1}) \right]$$

The maximum M_1 is

$$M_1 = \frac{\mu_1(2) + \mu_2(2)}{2} = 0.75$$

Thus the result of the combination is:

$$B = [(b_1, 0.73), (b_2, 1.0), (b_3, 0.73)]$$

The normed weighted sum combination yields using Eq. 7.1.6 the response set B:

$$B = \left[\left(b_1, \frac{\mu_1(1)/3.6 + \mu_2(1)/3.8}{(1/3.6 + 1/3.8)M_2}\right), \left(b_1, \frac{\mu_1(2)/3.6 + \mu_2(2)/3.8}{(1/3.6 + 1/3.8)M_2}\right), \right.$$

$$\left. \left(b_1, \frac{\mu_1(3)/3.6 + \mu_2(3)/3.8}{(1/3.6 + 1/3.8)M_2}\right) \right]$$

The maximum M_2 is

$$M_2 = \frac{\mu_1(2)/3.6 + \mu_2(2)/3.8}{(1/3.6 + 1/3.8)} = 0.75$$

Thus the result of the combination is:

$$B = [(b_1, 0.72), (b_2, 1.0), (b_3, 0.75)]$$

For the minimum and both additive cases, the resulting $b_{\max} = b_2$ was not preferred by any one of the rules.

7.2   Rule assessment

Rule assessment techniques described in Chapter 5 are mostly designed for systems with continuous response. In the case of discrete responses, different methods have to be used. An algorithm based on the idea used in the b-cut algorithm can be developed in this case. For each discrete response and every explanatory variable, one can find the range to which this response corresponds. If these "inverse images" are different for the different discrete responses, then the variable can be used as an argument in the rule system. In this case, the discrete and the continuous arguments should be treated separately. An algorithm to generate rules is the following:

1. In order to decide about the explanatory "power" of a continuous variable, its extremes in the training set have to be found:

$$a_{k\,\text{max}} = \max_S a_k(s)$$

and

$$a_{k\,\text{min}} = \min_S a_k(s)$$

2. Find the variable supports corresponding to the response supports:

$$\alpha_{i,k}^- = \min_S\{a_k(s) \text{ such that } b(s) = b_i\} \tag{7.2.1}$$

and

$$\alpha_{i,k}^+ = \max_S\{a_k(s) \text{ such that } b(s) = b_i\} \tag{7.2.2}$$

The membership value 1 is assigned to the mean value of the variable corresponding to the response support.

$$\alpha_{i,k}^1 = \frac{1}{N(i)} \sum_{b_i} a_k(s) \tag{7.2.3}$$

where $N(i)$ is the number of such elements s such that the response is

$$b_i$$

3. Calculate for each k the relative width of the smallest support:

$$\delta_i = \min_i \frac{\alpha_{i,k}^+ - \alpha_{i,k}^-}{a_{k\,\text{max}} - a_{k\,\text{min}}} \tag{7.2.4}$$

4. Variables with

$$\delta_i > \Delta$$

will not be considered as belonging to the rule system; the other ones will constitute the selected set of K explanatory variables. If for a variable the "inverse image" of all partitions is very wide, then it cannot be used as an explanatory variable. The value of Δ should depend on the number of rules to be found. (Good results were obtained with values in the approximate range 0.3-0.5.)

5. For discrete explanatory variables consider the complete inverse image of an element in the response set. Let $N_i(a_k^*)$ be the number of elements of the training set such that the k-th argument is a_k^* and the response is b_i. Formally:

$$N_i(a_k^*) = |\{a_k(s) = a_k^* \text{ such that } b(s) = b_i\}| \qquad (7.2.5)$$

6. Define the fuzzy set A_k^i as a fuzzy set on the possible (discrete) values of the argument a_k $(a_k^1, \ldots, a_k^L)$.

$$A_k^i = \left\{ (a_k^m, \frac{N_i(a_k^m)}{\max_l N_i(a_k^l)}), l = 1, \ldots L \right\} \qquad (7.2.6)$$

In this case the explanatory power of the argument a_k depends on the cardinality of A_k^i. A large cardinality means that all possible a_k values can yield b_i as a response - thus the argument is of little use. In contrast the smaller the cardinality the more one can conclude b_i or not b_i holds as a consequence of knowing the argument a_k. A threshold value can be set for the cardinality of A_k^i to be included into the rule system.

7.3   Application to weather classification

A concrete example dealing with the fuzzy rule-based classification of daily atmospheric circulation patterns is now presented. It has been known for many years that local or regional daily meteorological variables such as precipitation, temperature or wind depend on the type of atmospheric circulation pattern (CP) occurring over the region. Even daily weather forecasts in the newspapers or TV explain the possible next day weather from large scale features — cyclonic and anticyclonic fronts. The CP-s are classified air-pressure maps observed daily over a large area such as a whole continent. Past decade concerns about anthropogenic effects on the climate such as global warming have prompted several researchers to use modern system and statistical techniques to investigate this linkage. In this manner, the effect of anthropogenic actions, for example an increase of CO_2 concentrations in the atmosphere may be investigated in terms of changes in the time series of CP types. However, the first stumbling block of this approach is the non-availability of a good typology or classification of CP over most of the earth. Western Europe is a most remarkable exception, thanks to the long persistence of the meteorologists in the German Weather Service (DWD 1984) who developed an excellent subjective classification, described in Hess and Brezowsky (1969) and Bárdossy and Caspary (1990). The 29 types of CP, grouped into the three broad categories (or major types) of zonal, half meridional, and meridional are summarized in Table 7.1. FRB certainly has the capability to construct a good classification either on the basis of existing subjective classes or by itself.

7.3.1   Classifications approaches

CP classification techniques may be grouped into subjective or manual and objective or automated procedures (Yarnal, 1984, 1993). The former group has been in existence for over a century in Europe, as described in Baur et al. (1944), and about half a century in the US (Krick, 1944). The latter type has emerged as a result of the development of high-speed computers and availability of statistical software packages, mostly principal component analysis and clustering.

Objective or automated techniques, in contrast, are based on mathematical approaches such as hierarchical methods (Johnson, 1967), K-means methods (MacQueen, 1967), and correlation methods (Bradley et al., 1982; Yarnal, 1984, 1993). Thus, in an eastern Nebraska study, nine types of CP's have been identified after having performed a principal components analysis coupled with the K-means method (Matyasovszky et al. 1992).

A common problem with statistical clustering methods is that the resulting CP-s have to be interpreted by meteorologists and often fail to provide a good description of actual weather conditions. Therefore an objective technique which incorporates the knowledge of the meteorologist is in order.

CP-s are classified using observed air-pressure data on a regular grid over a selected area. Instead of using the pressure data, the anomalies (departures from the normal values) are used for the classification as they describe the weather conditions better. The observed pressure maps are available from the gridded data set of the National Meteorological Center (NMC), US. Figure 7.1 shows the locations of the gridpoints (or pixels).

Anomalies form a good basis for a classification. For this purpose first the mean air-pressure is subtracted from the observed daily data. Let $h(u_k, t)$ be the observed 700 hPa surface at location u_k and time t,

Table 7.1. Subjective classification of European atmospheric circulation patterns (modified after Hess and Brezowsky, 1969)

Group of circulation types			
Major types		Subtypes	Abbrev.
Zonal			
West	W	1. West anticyclonic	W_a
		2. West cyclonic	W_z
		3. Southern West	W_s
		4. Angleformed West	W_w
Half-Meridional			
Southwest	SW	5. Southwest anticyclonic	SW_a
		6. Southwest cyclonic	SW_z
Northwest	NW	7. Northwest anticyclonic	NW_a
		8. Northwest cyclonic	NW_z
Central European high	HM	9. Central European high	HM
		10. Central European ridge	BM
Central European low	TM	11. Central European low	TM
Meridional			
North	N	12. North anticyclonic	N_a
		13. North cyclonic	N_z
		14. North, Iceland high, anticyclonic	HN_a
		15. North, Iceland high, cyclonic	HN_z
		16. British Islands high	HB
		17. Central European Trough	T_rM
Northeast	NE	18. Northeast anticyclonic	NE_a
		19. Northeast cyclonic	NE_z
East	E	20. Fennoscandian high anticyclonic	HF_a
		21. Fennoscandian high cyclonic	HF_z
		22. Norwegian Sea -	HNF_a
		Fennoscandian high anticyclonic	
		23. Norwegian Sea -	HNF_z
		Fennoscandian high cyclonic	
Southeast	SE	24. Southeast anticyclonic	SE_a
		25. Southeast cyclonic	SE_z
South	S	26. South anticyclonic	S_a
		27. South cyclonic	S_z
		28. British Islands low	TB
		29. Western Europe Trough	T_rW

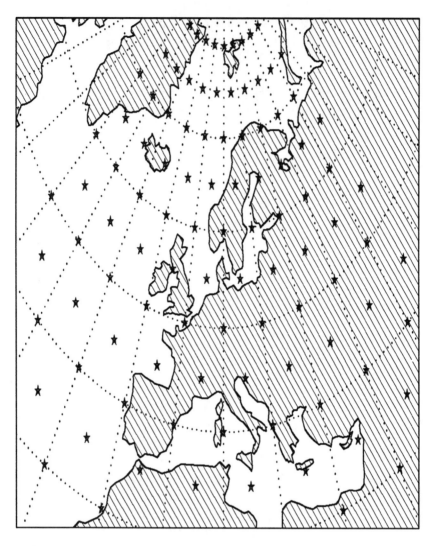

Figure 7.1.   The grid with air-pressure values used for classification

T be the total time horizon, and the temporal mean of $h(u_k, t)$ be:

$$\bar{h}(u_k) = \frac{1}{T} \sum_{t=1}^{T} h(u_k, t) \qquad (7.3.1)$$

In order to define the rules on the unit interval for every time t, the height is normalized using the formula:

$$g(u_k, t) = \frac{h(u_k, t) - \bar{h}(u_k) - \min_l \{h(u_l, t) - \bar{h}(u_l)\}}{\max_l \{h(u_l, t) - \bar{h}(u_l)\} - \min_l \{h(u_l, t) - \bar{h}(u_l)\}} \qquad (7.3.2)$$

The 700 hPa daily pressure surface is thus mapped on the interval $[0,1]$.

To classify CP's by the use of fuzzy rules, each CP type is first described by a set of rules and then the classification is done by selecting the CP type for which the DOF is the highest. This method corresponds to the case of maximum combination and maximum defuzzification. Because of the great number of pixels a rule would turn out to have 110 arguments. To reduce this set, rules are defined only for a subset of the pixels. This subset can be different for each rule. Each fuzzy rule corresponding to a circulation pattern of type i concerns a set of normalized pressure values (Eq. 7.3.2) at a few (2 to 5) selected pixels.

Four classes v of relative pressure values are defined. The membership functions are the portions with support $(0,1)$ of the TFN given below:

very low values, class $v = 1$: $(0, 0, 0.4)_T$

medium low values, class $v = 2$: $(-0.2, 0.2, 0.5)_T$

medium high values, class $v = 3$: $(0.5, 0.8, 1.2)_T$

very high values, class $v = 4$: $(0.6, 1, 1)_T$

A rule is defined by assigning specific classes to selected pixels. The combination of the corresponding membership values is done using the "MOST OF" and the "AT LEAST A FEW" operators described in Chapter 3. Thus a rule has the form:

If AT LEAST A FEW $(g(u_{i_1}, t)$-s $(i_1 \in I_1)$ are belonging to class $v=1$)
AND
MOST OF $(g(u_{i_2}, t)$-s $(i_2 \in I_2)$ are belonging to class $v=2$)
AND
MOST OF $(g(u_{i_3}, t)$-s $(i_3 \in I_3)$ are belonging to class $v=3$)
AND
AT LEAST A FEW $(g(u_{i_4}, t)$-s $(i_4 \in I_4)$ are belonging to class $v=4$)
THEN the CP on day t is i

Here the indices i_v indicate the gridpoints that should belong to class v according to the rule definition. The rules are defined by selecting the corresponding sets of indices I_1, T_2, I_3 and I_4. The AT LEAST A FEW operator is selected for classes 1 and 4 in order to ensure that a low (or a high) pressure center should be located somewhere on the selected gridpoints. The MOST OF is selected for classes 2 and 3 to exclude cases with undesired low (or high) pressures at the selected gridpoints.

The DOF of the above rules can be calculated using Equations 3.1.14 to 3.1.17 or 3.1.19 and 3.1.20 as:

$$\nu_i = D_1 D_2 D_3 D_4 \qquad (7.3.3)$$

where the D_v values are calculated as a convex combination of OR and AND functions:

$$D_v = \gamma F_o \left(\mu_v(g(u_{1_v})), \dots, \mu_v(g(u_{I_v})) \right) +$$

$$+ (1 - \gamma_v) F_a \left(\mu_v(g(u_{1_v})), \dots, \mu_v(g(u_{I_v})) \right) \qquad (7.3.4)$$

An alternative formulation for combining AND and OR DOF's is an l_p norm (see Section 3.1.3).

The verbal description of the circulation patterns given in Hess and Brezowsky (1969) was simply "translated" into fuzzy rules by selecting characteristic features of every CP. This classification scheme is applied to a measured sequence of daily 700 hPa elevations for the ten year period from 1977 to 1986.

The method is able to develop a semi-objective classification which resembles the subjective one and whose quality is measured by the differences between the conditional (conditioned on CPs) and the unconditional precipitation statistics, as described in Section 7.3.3.

Rules applicable to other regions have been defined by meteorologists possessing knowledge about local weather generating mechanisms. For this purpose, an interactive computer program was developed and used in a large number of geographical regions.

7.3.2   Numerical example

November 29, 1986, has been selected to illustrate the methodology. The CP type that persisted November 26-29, 1986, is of type "BM", according to the subjective classification. The fuzzy rule-based method also assigned the same classification. The membership functions of the four classes are given as above. The γ values depend on the CP, as they express the level of simultaneous fulfillment of a pressure value condition. A high γ value expresses the fact that even if a few pixels do not fulfill the condition for belonging to a given class; the corresponding overall DOF which would be zero with a pure "AND" is positive. In this example, a constant value $\gamma = 0.7$ gives results in agreement with experimental observations.

Normalized pixel values, membership grades for the selected day and rule defining the CP type BM are shown in Table 7.2. For example, in the case $v = 4$, very high relative pressure value, $F_o = 1.00$ $F_a = 0.6375$ and $D_{24} = 0.7 \cdot 1.00 + 0.3 \cdot 0.6375 = 0.8912$. The overall fulfillment grade of the rule is calculated by Eq. 7.3.4 as the maximum value $D_2 = 0.3944$. This example shows that even in the case when one pixel does not fulfill the prescribed pressure level as it is the case of pixel 5 for $v = 2$, the fuzzy rule can assign it to the selected case.

On the selected day, the DOF of CP type Wz is found to be 0 and that of type HM, 0.1247.

7.3.3   Use for precipitation modeling

In this investigation, the main purpose of a CP classification was its use for hydrologic modeling, in particular, simulation of precipitation performed as indicated in Bárdossy and Plate (1992) and Bogardi et al. (1994). Table 7.3 shows different precipitation statistics of Essen (Germany) for three selected circulation patterns. Note the differences between CP10 being a CP associated with dry weather and CP02 which is very wet.

To measure the quality of a classification for precipitation estimation or generation, three information measures increasing with the information content of the method are introduced.

A CP classification would be perfect for the description of precipitation occurrence if the conditional rainfall probabilities would be either close to 1 or 0. This would mean that the knowledge of the occurrence of a CP would with high probability imply wet or dry days, depending on the CP. The first information measures the predictability of wet and dry days depending on the CP only. For this purpose the sum of the squared difference I_1 between the conditional probability p_{A_t} of rainfall an day t, given that the CP is A_t and the unconditional probability p of rainfall at a given site is calculated:

Table 7.2. Numerical example of calculation of DOF for CP type BM that occurred on Nov. 29, 1986.

Rule Class	Pixel		Normalized	Membership	DOF
	Long.	Lat.	Pressure g	Value μ	D_{2v}
$v = 1$	W25^o	N65^o	0.13	0.68	
Very low	W15^o	N65^o	0.00	1.00	0.8265
	W0^o	N70^o	0.15	0.62	
	W15^o	N35^o	0.63	0.00	
$v = 2$	W25^o	N75^o	0.12	0.84	
Medium low	W0^o	N80^o	0.18	0.96	0.6983
	E25^o	N75^o	0.31	0.63	
	W20^o	N40^o	0.62	0.40	
$v = 3$	W15^o	N45^o	0.84	0.92	
Medium high	W0^o	N50^o	0.96	0.68	0.7473
	E10^o	N50^o	0.95	0.70	
	E15^o	N45^o	0.85	0.90	
$v = 4$	W5^o	N55^o	0.94	0.85	
Very high	E5^o	N55^o	1.00	1.00	0.8912
	E15^o	N55^o	0.90	0.75	

Table 7.3. Precipitation statistics for the station Essen (Germany) in Winter for a few selected circulation patterns.

Circulation Pattern	Freq. [%]	Prob. [%]	Contr. [%]	Rel.contr [-]	mean [mm]	std. [mm]
CP02	17.77	81.68	38.55	2.17	6.54	6.05
CP06	5.08	86.96	8.66	1.71	4.83	5.02
CP10	15.62	33.22	6.95	0.45	3.30	4.41

Freq. = circulation pattern occurrence frequency
Prob. = probability of a wet day for the given CP
Contr. = contribution of the pattern to the total winter precipitation
Rel.contr. = Contr./Freq.
Mean = mean daily precipitation amount on wet days for the given CP
std. = standard deviation of daily precipitation amounts on wet days for the given CP

$$I_1 = \left(\frac{1}{T-1}\sum_t (p_{A_t} - p)^2\right)^{\frac{1}{2}} \qquad (7.3.5)$$

Besides the predictive power of CP-s in forecasting wet and dry days, the different behavior of CP-s regarding precipitation amounts is also interesting. The closer the mean conditional precipitation amounts are to the observed values, the better is the classification. This aspect is measured by the second information content index I_2 which is equal to the sum of squared difference between conditional and unconditional rainfall means:

$$I_2 = \left(\frac{1}{T-1}\sum_t (m_{A_t} - m)^2\right)^{\frac{1}{2}} \qquad (7.3.6)$$

where m is the unconditional mean daily rainfall amount at a given site and m_{A_t} the mean daily rainfall conditioned on the CP being of type A_t. The third information content measure I_3 also depends on the mean rainfall m and measures the relative deviation from that mean value given the type of CP A_t that occurs on day t:

$$I_3 = \frac{1}{T} \sum_t \left| \frac{m_{A_t}}{m} - 1 \right| \tag{7.3.7}$$

The more the precipitation behavior differs from the mean, the higher I_3 is.

In order to compare the subjective and fuzzy classifications, the mean information content of 25 stations in the Ruhr catchment has been calculated for both the subjective classification of Hess and Brezowsky (1969) and the fuzzy rule based classification. As shown in Table 7.4 for summer and Table 7.5 for winter, the subjective classification seems to deliver the best results. However, the three measures of information content, I_1, I_2, and I_3 do not vary much between the three approaches, the difference between the two seasons being larger than that between the approaches. Furthermore a subjective classification is rarely available, and the Hess and Brezowsky (1969) classification is based on 111 years of data is unique in the world. In the case considered, the fuzzy rule-based approach yields somewhat better information content than the k-means technique.

Table 7.4. Average information content of three CP classification schemes (Summer).

Classification Method	I_1	I_2	I_3
Hess-Brezowsky	0.243	1.996	0.609
Fuzzy rules	0.211	1.778	0.521
k-means	0.167	1.321	0.422

Compared to other classification approaches, fuzzy rules offer a method that is fast and easy to use. It directly incorporates meteorological knowledge; thus, an interpretation of the classes takes place before rather than after the classification has been constructed.

Further details on the physical aspects of this example and its potential use in climatic uncertainty modeling may be found in Bárdossy et al. (1994). Another application of the same fuzzy rule-based classification technique to the western US, where only rudimentary subjective CP classification was available, has been developed in Ozelkan et al. (1994).

Table 7.5. Average information content of three CP classification schemes (Winter).

Classification Method	I_1	I_2	I_3
Hess-Brezowsky	0.265	2.559	0.679
Fuzzy rules	0.217	2.251	0.629
k-means	0.188	1.975	0.592

8

Application to time series

The goal of time series analysis is to find explanations for the regular and irregular changes in the series. The description of time dependent data is a classical problem usually treated by statistical methods.

Fuzzy rules can alternatively be used to describe the behavior of time series. Rules can be used to describe the next state of an observation depending on the previous states and on the value of other observations. This type of description of time series can well be used for explanation of time series changes and forecasting — but at the present moment it is not possible to generate time series with the desired statistical properties using the rule-based approach. It is often much simpler to state the regular features of a time series in the form of rules, than to find an appropriate stochastic model having the desired properties. Another advantage of this type of modeling is that imprecise data can also be taken into account — which is an extremely difficult task in classical time series analysis.

Rules for this purpose can be derived with methods described in Chapter 5. However, in forecasting, an adaptive rule assessment can improve the the performance of a fuzzy model. In this case, the rule system is regularly updated using the newest observations to account for possible slight changes in the system.

8.1 Rule assessment

8.1.1 Adaptive rule assessment

Suppose the time dependent training set $\mathcal{T}(T)$ consists of a (vector valued) time series of length T:

$$\mathcal{T}(T) = \{(a_1(t), \ldots, a_K(t), b(t)) \; ; t = 1, \ldots, T\} \qquad (8.1.1)$$

After each time step the new element corresponding to time step $T+1$ is added to the training set. As the internal dependencies might gradually change with time it is reasonable to update the rules after the arrival of a certain amount of data. This update should consider the new observations with a higher weight than the old ones. The simplest way to do so is that one only considers data from the last τ observations defining the training set at time T as:

$$\mathcal{T}(T) = \{(a_1(t), \ldots, a_K(t), b(t)) ; t = T - \tau, \ldots, T\} \tag{8.1.2}$$

This "moving window" ensures that only the most recent data are used for rule assessment. If one uses the least squares rule estimation procedure a t dependent weighting can also be considered. Instead of the "unweighted" squared differences as in Eq. 5.3.12 the t weighted ones are to be minimized

$$\sum_{t=1}^{T} \frac{t}{T} \left(\frac{\sum_{i=1}^{I} \nu_i(t) M(B_i)}{\sum_{i=1}^{I} \nu_i(t)} - b(t) \right)^2 \to \min \tag{8.1.3}$$

This leads to a linear equation system similar to Eq. 5.3.14:

$$\sum_{t=1}^{T} \frac{t}{T} \frac{\sum_{i=1}^{I} \nu_i(t) \nu_j(t) M(B_i)}{(\sum_{i=1}^{I} \nu_i(t))^2} = \sum_{t=1}^{T} \frac{t}{T} \frac{\nu_j(t) b(t)}{\sum_{i=1}^{I} \nu_i(t)} \tag{8.1.4}$$

There are I unknowns $M(B_1), \ldots, M(B_I)$ and I equations; thus, the above system usually provides the unknown quantities. Even though the equation system has to be solved at each time step t the system can be updated simply by considering only $a_k(T+1)$-s and $b(T+1)$.

8.1.2 Rule assessment from trajectories

Variables to be modeled are often observed as time series, and rules are to be found that describe the transition of the system from a previous state to the other. In this case the response corresponding to a time step is one of the arguments for the next time step. This property makes rule assessment extremely difficult, as small errors might also cause divergence.

Formally such a case can be described as follows: Suppose a time series of the form

$$(a_1(t), \ldots, a_I(t), b(t)) \quad \text{for} \quad t = 1, \ldots, T$$

is known and to be modeled. Furthermore suppose that the first argument is the output of the previous time step:

$$a_1(t) = b(t-1)$$

A rule system that describes $b(t)$ with the arguments $(a_1(t), \ldots, a_I(t))$ is sought. Let $R^{(t)}$ be the response time series at time t. Formally it is defined as:

$$R^{(1)} = R(a_1(1), \ldots, a_I(1)) \tag{8.1.5}$$

$$R^{(t)} = R(R^{(t-1)}, a_2(t), \ldots, a_I(t)) \tag{8.1.6}$$

Then a reasonable rule assessment strategy is to minimize the difference between the observed and the calculated trajectories:

$$\sum_{t=1}^{T} (R^{(t)} - b(t))^2 \longrightarrow \min \tag{8.1.7}$$

This minimization problem is usually a non-linear problem and has thus to be solved with appropriate optimization methods. The cases such as fuzzy valued time series $\hat{b}(t)$ and fuzzy valued arguments can be treated similarly as presented in Chapter 5.

8.2   Example: Water demand forecasting

The forecasting of water demand is a very important problem in reservoir operation. Unnecessary releases lead to water losses, but enough water should be released to prevent shortages or water quality problems.

Traditional ways to forecast water demand is to apply either standard regression techniques or time series modeling, like ARMA. In order to obtain reliable model parameters, long time series are required. However, this may result in long term changes often being neglected.

Water demand in an area depends on several variables including some parameters determining the long term behavior such as:

industrialization level of the area

population

agricultural activity

general economic output

Variables that determine mostly short range demand include:

temperature (daily mean and maximum): water use is usually higher on warm days

precipitation or air moisture: water use is usually higher on dry days

holiday — working day cycles: industrial water use is mostly concentrated over working days,

The short range factors can usually be handled by standard models but the long range parameters yield nonstationary time series, and make the forecasting based on past observations extremely difficult. Figure 8.1 shows the daily water demands of the year 1983 in the Ruhr catchment (Germany).

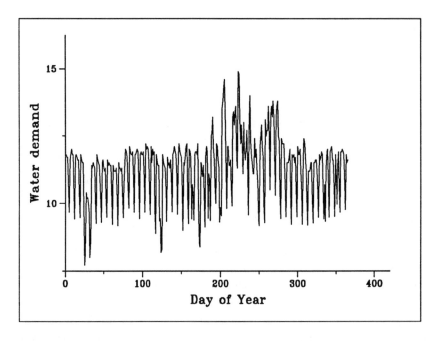

Figure 8.1.   Daily water demands for the hydrologic year 1983 (11.1.1982 – 10.31.1983) for the Ruhr catchment.

8.2.1   The model

After an analysis of the available data it was decided that the model should be based on three explanatory variables — the day of the week, the daily maximum temperature, and the general weather conditions of the previous days.

The day of the week plays a major role in water demand. Three types can be identified

1. Working days: on these days the water use is the highest due to industrial activities.

2. Saturdays: less water is used than on working days, but as some businesses still have working hours on Saturday the water demand is higher than on Sundays.

3. Sundays and public holidays: these days exhibit the lowest water demands.

One can observe the regular drop in water use on each weekend in Figure 8.1. This variable being purely deterministic, can be considered as a crisp argument. Rules for each of these classes are assessed separately.

Daily maximal temperature values $TX(t)$ are taken in C^0 as observed. It is used as a fuzzy argument in the rule system. The weather conditions on the previous days are described by the variable $GL(t)$. This variable is in our analysis conditioned on atmospheric circulation patterns (CP), which were divided into two subsets — the dry patterns and the wet patterns. The patterns used for this study correspond to the classification example of Chapter 7. The variable entering the analysis is the length (in days) of the interval with dry patterns. If the length exceeds 5 then it has no increased influence. It is also used as a fuzzy argument in the rule system.

Twenty-seven fuzzy rules are derived from a training set consisting of a one year record. Nine rules describe the behavior on working days, nine on Saturdays, and nine on Sundays. The rules are derived with the help of the counting algorithm. The rules are formulated both as verbal and fuzzy rules. For example:

If it is hot on a working day and the length of the previous dry CP interval is long, then the water use will be large.

This can be formulated as

If TX is $(20, 26, 35)_T$ $^\circ C$ and GL is $(3.5, 5, \infty)_T$ then

$$EV \text{ is } (5.38, 5.90, 6.58)_T \; m^3/s \qquad (8.2.1)$$

For the maximum daily temperature TX and the length of dry weather GL three fuzzy sets are defined and used in every possible combination — resulting in nine rules.

Three years of daily water use data at Villigst, Germany is analyzed. The same rule assessment procedure was applied for rule assessment using one year of data as a training set, and the other two as validation set for model testing. This procedure thus delivers three rule systems with corresponding validation measures.

To avoid possible bias caused by changes in the industrial use a multiplicative correction term is applied to the fuzzy rule based models (hybrid rule system) and the neural net mentioned below. This means:

$$W^*(t+1) = \frac{W(t)}{R\left(d(t), (TX(t), GL(t))\right)} R\left(d(t+1), (TX(t+1), GL(t+1))\right)$$

$$(8.2.2)$$

here $W(t)$ is the water demand on day t, $W^*(t)$ is the estimated water demand on day t. R denotes the rule response.

8.2.2   Comparison to other models

In a comparative study, three different models are used to describe the water loss in the selected region. They are as follows:

1. ARMA(1) process model

2. neural net model

3. fuzzy rule based model

To compare the results with other methods an ARMA(1,1) model is calibrated on the same data, and also a less conventional neural network model has been trained to model the time series. All three models use exactly the same input data. In the ARMA(1,1) approach the square of TX is used as an additional variable.

Table 8.1 shows a comparison of the results of the three different models. Five performance indices were selected to judge the performance of the three different approaches. These are the mean error (ME), the mean absolute error (MAE), the standard error (SE), the maximal error (MAX), and the correlation coefficient between the observed and calculated demand series (COR).

There does not seem to be a substantial difference between the three approaches when applied to the same data as for model identification. In the case a different data set is used there is a systematic bias for the ARMA model the mean error ME being non-zero. MAE and SE are usually the smallest for the fuzzy model (FRB). The maximum error MAX is somewhat larger for FRB than for NN and ARMA and correlation KOR slightly smaller. However, given the simplicity of the FRB model, we are ready to recommend it.

Table 8.1. tenrm Modeling errors for water use at Villigst applying the three models ARMA = ARMA(1,1), NN = Neural Net, FRB = Fuzzy rule-based model; the criteria are:

calibration year	calculated year	Model	ME [m3/s]	MAE [m3/s]	SE [m3/s]	MAX [m3/s]	COR [-]
		ARMA	0.00	0.17	0.26	2.20	0.93
	EV-76	NN	-0.01	0.20	0.27	1.00	0.93
		FRB	0.00	0.19	0.30	2.29	0.92
		ARMA	-0.25	0.30	0.37	1.30	0.92
1976	EV-83	NN	-0.02	0.24	0.41	2.50	0.81
		FRB	0.00	0.20	0.28	1.83	0.90
		ARMA	-0.31	0.36	0.44	1.40	0.86
	EV-90	NN	0.00	0.21	0.27	1.00	0.87
		FRB	0.00	0.27	0.36	1.49	0.79
		ARMA	0.25	0.31	0.39	2.00	0.92
	EV-76	NN	0.00	0.24	0.35	2.10	0.88
		FRB	0.00	0.21	0.33	2.33	0.89
		ARMA	0.00	0.14	0.20	1.00	0.93
1983	EV-83	NN	0.00	0.15	0.20	0.90	0.93
		FRB	0.00	0.17	0.25	1.40	0.91
		ARMA	-0.06	0.20	0.25	0.90	0.89
	EV-90	NN	0.00	0.20	0.28	1.10	0.86
		FRB	0.00	0.22	0.30	1.30	0.83

ME = Mean error
MAE = mean absolute error
SE = Standard error
MAX = Maximum error
COR = Correlation coefficient

8.2.3 Application to forecasting

The previously described model only served to find an explanation of the water demand as a function of other observed parameters. In practice reservoir control requires decisions a few days in advance, such as the flow time between the release and the location where the demand occurs usually is a few days. Therefore, inexact daily maximum temperatures and circulation patterns have to be considered for the forecast of the water demand. There are two possibilities — either to build up the forecasting model on past observations only, or to use the weather forecasts. The second approach uses more information and thus is the reasonable solution. However, weather forecasts are imprecise and this imprecision has to be taken into account. This can be done by taking the forecasted temperatures as fuzzy numbers. These fuzzy numbers can be assessed comparing past forecasts to the corresponding observations. Past temperature forecasts were grouped according to their verbal form:

1. above TX degrees

2. around TX degrees

3. under TX degrees

For each type of forecast, the difference between the forecasted TX_f and the observed TX_o temperature is calculated over a time horizon. These differences are used to assess the forecasted fuzzy temperature. Triangular fuzzy numbers are assessed, the mean difference receiving 1 membership, and mean $\pm$ 2 times standard deviation of the differences forming the two other points of the triangle. This fuzzy premise is then used in the rule system. The fulfillment grades are calculated using Eq. 3.4.1, and a crisp forecast is determined using the normed weighted sum combination and the mean defuzzification method.

Figure 8.2 shows the scattergram of the observed and the forecasted (5-day) water demand. Note the good quality of the forecasts even using imprecise weather forecasts as input variables.

8.3   Example: Daily mean temperature

The mean daily temperatures show a marked annual cycle. The mean temperature on a particular day at a particular site of the year depends on a set of variables:

the season (or month)

the precipitation

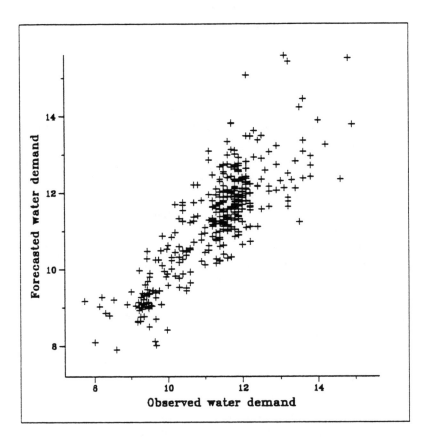

Figure 8.2. Observed versus 5 day forecasted water demand using the fuzzy rule based model.

the atmospheric circulation

the temperature on the previous day

There are models that describe this process using stochastic methods (Richardson 1981). There is also a possibility to describe them using fuzzy rules. In this section, only a part of this modeling is described, namely, the mean daily temperature as a function of the day of the year.

The problem here is to find a periodic function describing the annual cycle of daily mean temperatures. This is usually done by using long temperature records and fitting them with Fourier series. Rules can also be used for this purpose; alternatively even shorter series can be used. As a single rule argument the day of the year (Julian date) is used. Rules are set up on a monthly basis with an overlap of 30 days (the last 15 days of the previous month and the first 15 days of the next month). This way 12 rules are considered. For example, for April:

$$(74, 105, 135)_T$$

To ensure the periodicity for the days it is assumed that day 1 follows day 365. Therefore the rules for January and December are not proper fuzzy numbers but this does not affect the approach. After having defined the rule arguments the least squares algorithm was used to assess the rule responses — based on ten years observation data at a particular site in Germany (at the city of Trier). Figure 8.3 shows the rule response function thus obtained. As a comparison the mean temperatures calculated with the help of a 10 day moving average over the 10 year is also drawn. Note the good agreement between the two curves, but the rule-based one is smoother and without any errors due to the relatively short time period. The other factors mentioned above can also be incorporated into a revised temperature model. In order to reduce the number of rules, rule responses can be formulated individually, for example in the form:

In summer if there is rainfall on a given day then the mean daily temperature is lower than normal.

The quantitative form of these rules can be assessed individually, but using the output of the normal daily temperature rules. Then, on a particular day, one can calculate the normal daily temperature using the above sketched rule system. This value is then modified by the successive rules. Thus, by knowing the daily precipitation and the Julian date, rule responses can be calculated.

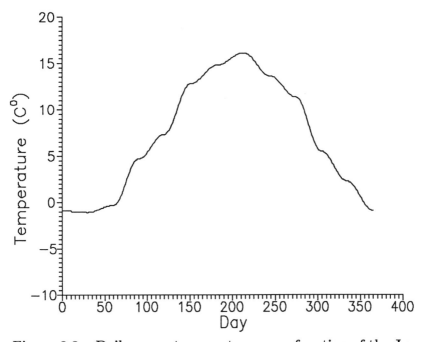

Figure 8.3.   Daily mean temperatures as a function of the Julian date, solid line = Rule response function, dashed line 10 days moving average (both calculated using 10 years observations at Trier (Germany).

9

Application to dynamical physical systems

This chapter is designed to demonstrate the applicability of fuzzy rule based modeling to a well-defined physical system traditionally described by partial differential equations. Differential equations (partial or ordinary) describe the relationship between the value of a function at a selected point as well as the neighboring points. Therefore such equations may also be regarded as a kind of functional description of a rule; for example the conservation of momentum equation is usually stated as: "The force applied is equal to the rate of change of momentum".

In practice there are seldom cases when a partial differential equation (p.d.e.) system can be solved analytically. Instead of the unknown exact solution numerical methods are used to find a good approximation. However an approximation can be obtained not only by a numerical solution of the p.d.e. using an appropriate discretization, but by replacing the pde. itself by rules. This approach can be very useful for applications in natural systems like meteorology — groundwater hydraulics or environmental pollutant transport where the initial and boundary conditions are usually only inexactly known. Furthermore, the effect of natural heterogeneity and variability of the parameters used in the p.d.e. on the investigated process are often unknown. Thus, the accuracy obtained by a numerical solution is not realistic — the difference between observations and model calculations is often found to be much greater.

The technique for transformating a p.d.e. into rules to describe dynamical systems depends on the type of system considered. Our experience shows that expert knowledge is essential for this step to identify the appropriate arguments and to construct rules that describe the main features of the process. In this chapter, the methodology is described by means of an example, namely, the water movement in the unsaturated soil (Bárdossy and Disse, 1993). The problem of heat transfer through a heterogeneous medium can also be treated similarly. Details on that problem are not given here due to space limitations.

9.1   Application to soil water movement

Recent environmental concerns have again directed the attention of hydrologists to the problem of infiltration and water movement in the unsaturated zone; This portion of the soil near the surface, also called vadose zone contains varying amounts of moisture mixed with gaseous substances, in contrast with the saturated zone, where most of the gases are dissolved in water. The vadose zone is an extremely important component of the water cycle in the ground, as its behavior determines aquifer recharge, pollutant transport, salt leaching, and plant growth. In addition to classical differential equation approaches, new methods were developed to describe unsaturated flow.

The main force acting on water in the soil is gravity. Besides this, mostly due to the surface tension of water, so-called capillary forces also play an important role for the water movement in the unsaturated zone. The interaction of these two forces depends both on the soil characteristics and on the actual state of the soil (water content). The initial and boundary conditions, like rainfall or irrigation penetrating through the surface, exhibit high spatial and temporal variability and introduce further difficulty for this problem.

There have been several attempts to describe the soil water movement mathematically. Up to now, the Richards equation (Richards, 1931) has been the most common basic mathematical expression for describing unsaturated flow phenomena. This equation models unsteady flow in a multidimensional anisotropic and nonhomogeneous soil matrix by means of a partial differential equation.

The combination of Darcy's law (v = vertical flow rate)

$$v = -K(\theta)\left(\frac{\partial \psi}{\partial z} - 1\right) \tag{9.1.1}$$

with the continuity equation

$$\frac{\partial \theta}{\partial t} = -\frac{\partial v}{\partial z} \tag{9.1.2}$$

yields the non-linear Richards equation:

$$\frac{\partial \theta}{\partial t} = \frac{\partial}{\partial z}\left[K(\theta)\left(\frac{\partial \psi}{\partial z} - 1\right)\right] \tag{9.1.3}$$

where
θ = moisture content

$K(\theta)$ = hydraulic conductivity (a function of the moisture content)
$\psi(\theta)$ = the matrix head (a function of the moisture content)
As there is a one to one correspondence between θ and ψ, usually the hydraulic conductivity is written as $K(\psi)$.

The shapes of the $\psi(\theta)$ and $K(\psi)$ curves can be described by the Van Genuchten-equations (Van Genuchten, 1980):

$$K_r(\psi) = \frac{\left(1 - \left(\frac{\psi}{h_\alpha}\right)^{n-1}\left[1 + \left(\frac{\psi}{h_\alpha}\right)^n\right]^{-m}\right)^2}{\left[1 + \left(\frac{\psi}{h_\alpha}\right)^n\right]^{\frac{m}{2}}} \tag{9.1.4}$$

$$\psi(\theta) = h_\alpha\left[\left(\frac{\theta - \theta_r}{\theta_s - \theta_r}\right)^{-\frac{1}{m}} - 1\right]^{\frac{1}{n}} \tag{9.1.5}$$

with $m = 1 - \frac{1}{n}$
n = Van Genuchten shape parameter (-)
$h_\alpha = 1/\alpha$ (cm)
α = Van Genuchten scale parameter (1/cm)
$pF_\alpha = \log(h_\alpha)$ (-)
θ_r = residual water content (Vol.- %)
θ_s = saturated water content (Vol.- %)

Using the above equations, five parameters $\theta_r, \theta_s, K_s, n, \alpha$ have to be specified for each spatial discretization unit.

For the modeling of water dynamics in the unsaturated zone, one has to solve the p.d.e. using suitable algorithms. The models can be grouped into analytical and numerical approaches, with the latter being far more popular. Analytical solutions are often more difficult to obtain because the coefficients of Richards equation are functions of the dependent variables.

Much emphasis has been directed to numerical solutions of the Richards equation. Various algorithms are used including the finite difference and the finite element methods.

A basic problem in applying these algorithms is that they require a large number of parameters which are only available at a few sites. In addition several of these parameters influence the models in a highly non-linear manner, so that results can be very sensitive to parameter changes. It is extremely difficult to estimate these parameters at unsampled locations. As a consequence the application of these models to real life cases is presently limited.

The main idea of using fuzzy rules is that the water movement at a certain time and at a selected point depends to a very high degree only on the conditions (moisture content) in the immediate neighborhood of the point. The simplest way to create a rule system for this problem would be to take all five above parameters as arguments. However this would not help in the problem of parameter estimation. Therefore a simpler model with less parameters was sought.

The fuzzy rules to describe the vertical water movement were formulated with the help of relative soil moisture contents. In order to simplify the model, the parameter set was reduced to the two linear coefficients, K_s and θ_s with the idea of including all the nonlinearities into the rule system. Let a rule be formulated verbally as:

If the relative soil moisture in element i is $A_{1,i}$ and the relative moisture content of the adjacent element below is $A_{2,i}$ then the normalized flux between the elements is B_i.

In this case, normalized flux means that a specific value K_s is selected for which the rules are specified. As the flux is proportional to the K_s value a simple multiplication makes the rule applicable to the case of a value of K_s different from the selected reference value. This means that the rule system is a hybrid one (section).

The rules for the flows can be assessed from experiments or by considering numerical solutions of the pde. The first approach would fully disregard the physical knowledge (expressed in the Richards equation). Furthermore measurement errors and inaccuracies would influence the rule system. Using numerical solutions of the Richards equation has the advantage that rules for arbitrary conditions (from very dry to very wet) can be assessed, by specifying appropriate initial conditions for the numerical solution.

A training set containing the output of different solutions corresponding to intense rainfall was used for calibrating the rule model.

From this training set, different rules were derived for fuzzy models with different spatial resolutions (2.5 to 15 cm). The rule assessment was performed using the counting algorithm (Section 5.3.2). The specification of the intervals for the arguments is extremely important in this case, as flow rates are highly variable in the case of high relative moisture contents, and steadier for dry soils. Therefore a uniform partition of the argument space is not useful. Several different rule systems were assessed and validated. The number of rules ranged from 20 to 100, yielding different accuracy. In order to simplify the calculations the normed weighted combination method and the mean defuzzification were used.

The steps in applying the Richards equation based fuzzy model are:

1. The actual moisture content θ is converted to relative moisture content:

$$\Theta = \frac{\theta}{\theta_s}$$

2. The actual relative moisture contents of two adjacent layers Θ_j and Θ_{j+1} are used to calculate the DOF grade ν_i for each rule i.

3. The flow $Q_{j,j+1}$ between the layers is obtained with the help of the normed weighted sum combination and the mean defuzzification. This leads to the flow:

$$Q_{j,j+1} = \frac{\sum_i \nu_i q_i}{\sum_i \nu_i} \frac{K_s}{K_s^*} \qquad (9.1.6)$$

where $q_i = M(B_i)$.

4. The actual flow is converted into a corresponding average moisture content of the layer.

5. Steps 1–4 are repeated for each time step.

In order to ascertain if such a model gives a good approximation of the solution of the Richards equation a great number of comparisons were done. Both analytical solutions (for selected particular cases where such solutions exist) and numerical solutions were considered for this purpose. Figure 9.1 shows the soil moisture calculated with a finite element solution of the Richards equation and the corresponding solution obtained with the fuzzy rule based model. Both models were run with the same initial and boundary conditions. The profiles correspond to the case of an intense rainfall applied to a homogeneous soil column after 0.25, 1.00 and 1.75 hours. One can see that the fuzzy model reproduces the profile quite well, there is practically no visible difference between the two solutions.

Figure 9.2 shows the difference between the finite element and the fuzzy solution as a function of simulation time. One can see that the mean difference never exceeds 0.3% which is much lower than the usual measurement error of soil moisture content.

9.1.1   Application

A good agreement between a theoretical model (Richards equation) and the fuzzy rule-based version does not mean that the fuzzy model reproduces reality reasonably well. To demonstrate this the model was compared to a great number of field experiment data.

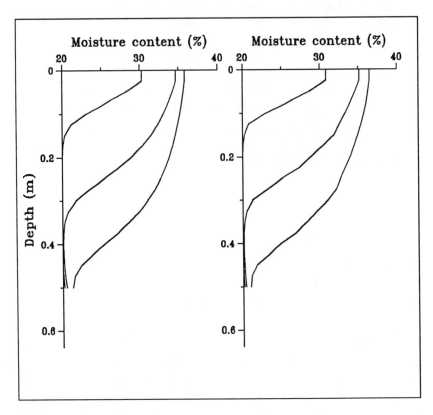

Figure 9.1. Soil moisture profiles obtained using finite element solution of the Richards equation (left) and the fuzzy rule based solution (right).
The profiles correspond to 0.25, 1.00 and 1.75 hours simulation time.

Data used to test the fuzzy model were collected within the scope of an investigation project which examined the infiltration behavior in a small area. The field had a length of 90 m, was grown with short cut grass and its soil was layered (loam, sandy loam, loamy sand). Altogether 20 locations were irrigated several times by a sprinkler infiltrometer. The infiltrometer could be adjusted to various intensities (from 5 up to 150 mm/h), and the rain energy represented by the drop size distribution was similar to natural rain. The infiltration plots (1 square meter each) were bounded by two circular sheets of metal pieces which were partly pressed into the soil. The overland flow was measured digitally, when the sprinkler irrigation exceeded the infiltration capacity. The impulse of a

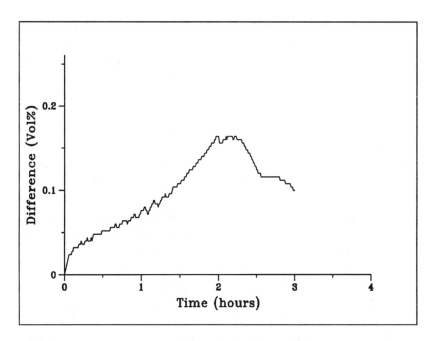

Figure 9.2. Mean absolute difference between the finite element solution of the 1 dimensional Richards equation and the fuzzy rule-based solution as a function of simulation time

small pump was stored on a personal computer. By knowing the exact pumping volume, this signal measured a mean discharge. In the same manner the rainfall intensity was evaluated. The soil moisture content was measured by Time Domain Reflectometry (TDR) at five different depths. The experiments were carried out in two distinct ways.

In Variant A the sprinkling rate was higher than the infiltration capacity. The initial water content at the five depths was measured and the infiltration rate was calculated by the difference between rainfall intensity and runoff.

In variant B the sprinkler irrigation was below the infiltration capacity (no overland flow). In this case both the rainfall intensity and the soil moisture content were measured throughout the experiment.

There are several simplified methods for the calculation of infiltration and soil moisture movement in the unsaturated zone. A widely used method was developed by Green and Ampt (1911). The fuzzy model was compared to a traditional Green & Ampt-based model using 25 soil moisture profile measurement curves (for 4 layers), and 32 runoff experiments. In each case model parameters were chosen so as to minimize the squared difference between the measured and calculated curves. Note that the applied rule system was not altered. It was assumed that each 15 cm layer is homogeneous and could be characterized by the same set of parameters. This means that for the Green & Ampt model each layer had 5 parameters: $K_s, \theta_s, \theta_r, n$ and α. The layers in the fuzzy model were described by two parameters K_s and θ_s. For the calculation of the fuzzy solution of the Richards equation, a 3 cm discretization was selected. In order to compare the results with the Green & Ampt model, 15 cm layers were considered to be homogeneous. This way each 15 cm layer consists of five 3 cm layers having the same K_s and θ_s values.

The results produced by the different models were compared to the measured curve s. Figure 9.3 shows a comparison of measured and calculated runoff for a selected experiment. One can see that the fuzzy Richards solution is very similar to the measured runoff. Note that the fit was obtained with only 40% of the parameters of the Green & Ampt model. Figure 9.3 shows a comparison of measured and calculated soil moisture contents in the different layers. In this case, as for runoff, the fuzzy Richards model gave the better results. For each layer the mean squared difference and the correlation coefficient between the measured and calculated curves were calculated. The same measures were also used for the comparison of the runoff experiments. Table 9.1 shows values of these performance measures. The fuzzy Richards model is superior to the other one.

Table 9.1. Comparison of measured soil moisture content curves and measured runoff with the results of the different models

Model	Soil Profile					Runoff	
	Squared error Vol. $(\%)^2$	Layer 1 ρ	Layer 2 ρ	Layer 3 ρ	Layer 4 ρ	Squared error $(mm/h)^2$	ρ
G & A Model	2.8	0.80	0.94	0.66	0.51	43.1	0.70
Fuzzy rules	1.2	0.93	0.96	0.83	0.78	24.8	0.80

9.1.2   Extension to two and three dimensions

The Richards equations can also be applied to horizontal water movement with the only difference being that the gravitational term has to be omitted. The traditional two and three dimensional finite difference or finite element solutions require numerical skill and high computational efforts. The one dimensional fuzzy model can easily be extended to two or three dimensions. Besides the already derived vertical rules, horizontal ones have to be derived. The one dimensional horizontal Richards equation (Eq. 9.1.3) without the gravitational term is written as:

$$\frac{\partial \theta}{\partial t} = \frac{\partial}{\partial z}\left[K(\theta)\frac{\partial \psi}{\partial z}\right] \qquad (9.1.7)$$

Solving the one dimensional Richards equation with different sets of initial and boundary conditions yields the training set. From this training set the rules for the horizontal flow are derived using the algorithm described in Section 5.3.2.

As in the case of vertical rules, the horizontal ones are based on relative soil moisture values. The form of a rule is:

If the relative soil moisture in element is $A_{1,i}$ and the difference between the relative moisture content of the element and the adjacent element is $A_{2,i}$ then the normalized flux between the elements is B_i.

Once these rules are specified, water movement is calculated as follows:

1. The selected region is discretized in accordance with the size of elements for which the horizontal and vertical rules were assessed.

2. The actual moisture content θ of a selected element is converted to relative moisture content:

$$\Theta = \frac{\theta}{\theta_s}$$

3. The actual relative moisture contents of the two vertically adjacent elements Θ_u and Θ_d are used to calculate the DOF grade ν_i for each vertical rule i.

4. The vertical flow for the element is then calculated using the normed weighted sum combination and the mean defuzzification.

5. The actual relative moisture contents of the four horizontally adjacent elements are used to calculate the DOF grade ν_i for each horizontal rule i.

6. The horizontal flow in the element is then calculated using the normed weighted sum combination and the mean defuzzification.

7. Steps 2–6 are repeated for each element.

8. The actual flows are converted into a corresponding average moisture contents of the elements.

9. Steps 2–8 are repeated for each time step.

Note that the required computational effort is a linear function of the number of elements considered and it is proportional to the number of rules used. The rule system should be selected according to the desired accuracy.

9.1.3   Comparison to numerical solutions

In order to check whether or not the fuzzy rule-based solution delivers results similar to the finite element method, a few examples were calculated with both methods.

Figure 9.3 shows a comparison of the finite element and the fuzzy rule-based solutions. One can see that the agreement between the two solutions is quite good.

Figure 9.4 shows the mean absolute difference between the finite element solution of the two dimensional Richards equation and the fuzzy rule based solution as a function of simulation time. Other examples also show that the mean deviation between the models at any time step is less than 1% (Vol.). This quantity is lower than the usual error in soil moisture measurements.

9.1.4   Discussion

The fuzzy rule based model for unsaturated flow has several advantages, namely:

model calculations require little computer time compared to the finite difference or the finite element solution

the corresponding computer code remains transparent, removing another frequent source of error

the difference between the fuzzy rule based and the finite difference or finite element solutions of the Richards equation are less than the usual measurement error.

However there are some disadvantages, such as:

rules have to be specified for each soil type

space and time discretizations cannot be selected in an arbitrary way

These problems could be overcome by creating a rule bank with the help of a great number of p.d.e. solutions.

FEM solution

Fuzzy solution

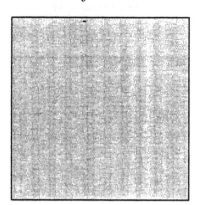

Initial state

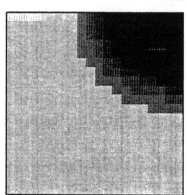
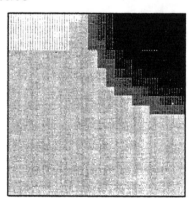

After 1 hour

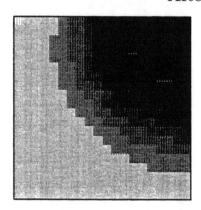
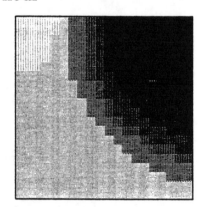

After 2 hours

Figure 9.3. A comparison of the solutions of an infiltration problem (watering part in a homogeneous soil)

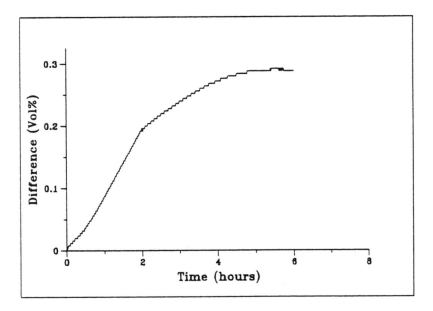

Figure 9.4.   Mean absolute difference between the finite element solution of the two dimensional Richards equation and the fuzzy rule based solution as a function of simulation time

Other applications

Besides the ones already presented, there are a great number of possibilities to apply fuzzy rules for modeling. In this chapter, an application to aid medical decision-making (the diagnosis of the severity of hypertension) is presented. The second example presents a sustainable reservoir operation rule by using a fuzzy rule-based model, which is simple, accurate, and representative of the system and is illustrated by a case study.

10.1   Application to medical diagnosis

The purpose of this chapter is to present a method of combining and reproducing the opinions of multiple medical experts by means of a fuzzy rule-based approach. A detailed description of the application can be found in Blinowska et al. (1992). In medical sciences experts gather empirical knowledge and experience in diagnosis and treatment of diseases. There are no generally accepted strict laws expressed in precise mathematical form as in "hard disciplines" such as physics. This kind of "soft" discipline provides ideal areas of application of fuzzy methods.

The fuzzy rule-based approach is applied to arterial hypertension description and includes questions related to disease severity, indication of etiological check-up, hospitalization, coronary risk, and indication of antihypertensive treatment. These are questions that a physician may have to answer regarding any hypertensive patient. Five medical experts in the hypertension field have provided their answers to five questions for one hundred patient files, case by case.

The stated aim of reproducing the (average) opinion of five experts has been reached, and extremely encouraging results have been obtained.

The methodology consists of the following steps:

1) construction of the questionnaire

2) information acquisition

3) target definition

4) selection of evaluation and internal consistency criteria

5) membership function construction and discriminant parameter selection

6) fuzzy rules construction and prior sensitivity analysis

7) application of the fuzzy rules to the validation or test set

10.1.1   Construction of the questionnaire

The questionnaire consisted of five initial questions, each with three coded possible answers, as follows:

1) severity of hypertension: mild (1), moderate (2), severe (3)

2) indication of antihypertensive treatment: no (-1), I do not know (0), yes (1)

3) indication for high blood pressure etiological check-up: no (-1), I do not know (0), yes (1)

4) indication for hospitalization: no (-1), I do not know (0), yes (1)

5) estimation of coronary risk: low (1), moderate (2), high (3)

The answer "I do not know" to questions 2, 3 and 5 reflects the expert's hesitation between "yes" and "no" answers. This is often the case in medical categorization procedures, which are particularly difficult to model with crisp techniques. Whenever a physician has to describe a continuous reality in terms of discrete linguistic categories, s(he) may be tempted to use one or more intermediate categories. No precise and well agreed upon boundary may exist between "yes" and "no".

A list of experimental data and patient characteristics considered as necessary and sufficient to answer the set of questions has then been established by medical experts. This list of potential premises consists of 14 clinical history data and 22 data gathered during the examination, as shown in Table 10.1. These data are selected in accordance with the medical experts' opinion. Only routinely collected data were considered.

10.1.2   Information Acquisition

A randomly selected sample of a hundred patient files has been submitted to five Hospital Broussais (Paris, France) experts in the hypertension field who were requested to follow their regular practice of examining the complete file of a patient and making a decision. Each expert gave his answers independently, without the knowledge of other experts' answers, according to his interpretation of the meaning of the questions.

Table 10.1.   Preliminary list of 14 clinical history and 22 examination data that medical experts consider necessary and sufficient premises to answer questions 1 to 5.

1. systolic blood pressure at discovery of hypertension (mm Hg)

2. maximum systolic blood pressure in past history (mm Hg)

3. hypertension duration (years) regular tobacco consumption: if yes

4. duration of tobacco consumption (years)

5. number of cigarettes per day

6. history of cardiac failure (no=0, yes=1)

7. history of coronary disease (no=0, yes-1)

8. arteritis of lower limbs (no=0, yes=1)

9. confirmed carotid stenosis (no=0, yes=1)

10. history of urinary tract infection (no=0, yes=1)

11. history of diabetes (no=0, yes=1)

12. history of high blood pressure in parents (0,1,2)

13. history of stroke in parent (0,1,2)

14. history of coronary failure in parent (0,1,2)

data gathered during the examination:

1. sex

2. age (years)

3. height (m)

4. bodyweight (kg)

5. tobacco consumption: if yes, number of cigarettes per day

6. consumption of alcohol (number of drinks per day)

7. headaches (no=0, yes=1)

8. sweating (no=0, yes=1)

9. palpitations (no=0, yes=1)

10. systolic blood pressure lying (mm Hg)

Table 10.1 **data gathered during the examination:** – contd.

11. diastolic blood pressure lying (mm Hg)

12. carotid murmur (no=0, yes=1)

13. femoral murmur (no=0, yes=1)

14. lumbar murmur (no=0, yes=1)

15. Sokoloff (mm)

16. repolarization abnormalities on EKG (no=0, yes=1)

17. creatininemia (μmol/l)

18. total cholesterol (mmol/l)

19. fasting blood glucose (mmol/l)

20. uricemia (μmol/l)

21. kaliemia (mmol/l)

22. hematocrit (%)

In order to define a proper target for the decision aid, the internal coherence of experts' opinions is first studied. Histograms of maximum differences between the five experts' coded responses to each question are sketched in Fig. 10.1. A difference of "zero" is ideal, "one" is acceptable and "two" means complete disagreement. The histograms corresponding to question 1 and, to a lesser extent, to questions 2 and 5, seem to present reasonable dispersion. In contrast, the histograms corresponding to questions 3 and 4 reflect considerable and rather astonishing vagueness so that one may wonder if all experts are interpreting these two questions in the same way. Further steps of our analysis will be illustrated by the examples of the first and second questions only, corresponding to severity of hypertension and indication for hypertensive treatment.

10.1.3 Target Definition

Since we consider that all experts' opinions are of equal value, the answer to question $j = 1, 2$ for patient i averaged over the five experts $r = 1, \ldots, 5$ $A_i^*(j)$, seems to be an appropriate and natural target:

$$A_i^*(j) = \frac{1}{5} \sum_{r=1}^{5} A_i(j, r) \qquad (10.1.1)$$

where $A_i(j,r)$ is the rth expert's answer to question j for patient i. Alternative targets may be weighted answers, using crisp or fuzzy weights, as in Bárdossy et al. (1993a) or else, a percentile of the answers; the procedure would remain the same as here.

Note that even though the answers to the questions are discrete, a rule system with a continuous type target is used. The reason for this is that the answers are discrete, but they are naturally ordered. This means a yes and a no answer yields as a consequence "I do not know". The intermediate values represent a varying degree of certainty of a response to the question under consideration.

10.1.4   Selection of Method Evaluation and Internal Consistency Criteria

Following proper scientific procedure, the data are split in a random way into two sets of $R = 50$ patient files each, the training set and the validation set. In order to evaluate both the dispersion of experts answers and model answers, two criteria are used:

(1) the sum of the squared differences $S^2(j,r)$, giving a measure of the variance of the rule consequence:

$$S^2(j,r) = \sum_{i=1}^{50}(A_i^*(j) - A_i(j,r))^2 \qquad (10.1.2)$$

where r=0 corresponds to model output and r=1,...,5 to experts' responses.

(2) The other criterion is the sum of the differences $D(j,r)$, giving a bias measure

$$D(j,r) = \sum_{i=1}^{50}(A_i^*(j) - A_i(j,r)) \qquad (10.1.3)$$

where r=0 corresponds to model output and r=1,...,5 to experts' responses.

10.1.5   Membership Function Construction and Discriminant Parameter Selection

Using the calibration set, patients are categorized according to the average coded opinion of the experts, rounded off to the closest integer.

For each experimental parameter n and question j, triangular membership functions $\mu_{n,k}^{j}(u)$ are assessed using the counting algorithm (Section 5.3.2) applied as if only the parameter under consideration were available. This way, for each parameter a univariate rule system consisting of three rules corresponding to each category $k(k = 1, 2, 3$ or $k = -1, 0, 1)$ is constructed. In order to evaluate the nth parameter discriminant power, a partial category index function PCI is defined as the rule response function corresponding to the selected experimental parameter n and question j.

$$PCI_n^{j}(u) = \frac{\sum_k k \mu_{n,k}^{j}(u)}{\sum_k \mu_{n,k}^{j}(u)} \qquad (10.1.4)$$

Three examples of $PCI_n^{j}(u)$ variation are given in Fig. 10.2. For "CDBP" (diastolic blood pressure corrected for age), the partial category index varies monotonically from 1 to 3; for the "Sokoloff" it also varies monotonically from about 1.5 to 3: in other words, it never indicates the first category but stays within a narrower range. For the "creatinine" it is not a monotonic function. At this first stage of parameters selection, all parameters with non-monotonical partial category indices are eliminated, which is a generally recommended procedure.

Membership functions of binary parameters taking on values of yes and no only are constructed in the following way:

Let $f_{nk}^{j}(\text{yes})$ and $f_{nk}^{j}(\text{no})$ be respectively the frequencies of occurrence of "yes" and "no" answers to the question j about parameter n within the k-th category. In case of a combination of variables, such as parental antecedents, the answer is taken as "no" if there is no antecedent and "yes" if there is at least one.

$$\mu_{nk}^{j}(\text{yes}) = \frac{f_{nk}^{j}(\text{yes})}{f_{nk}^{j}(\text{yes}) + f_{nk}^{j}(\text{no})}$$

and

$$\mu_{nk}^{j}(\text{no}) = \frac{f_{nk}^{j}(\text{no})}{f_{nk}^{j}(\text{yes}) + f_{nk}^{j}(\text{no})} \qquad (10.1.5)$$

These assessments are performed for each question j.

Table 10.2 shows the parameters that have been retained for further analysis. Note that systolic and diastolic blood pressure values have been corrected for patient age, as explained in Blinowska and Duckstein (1993).

Table 10.2. Severity of hypertension and indication for hypertensive treatment parameters retained for further analysis and their mean values (mean), lower limits (L-), upper limits (L+) within each category, or, for binary parameters, f(yes), i.e., the ratio of the number of "yes" values divided by the total number of patients within a given category.

Severity	category 1 (32 patients)			category 2 (12 patients)			category 3 (6 patients)		
parameter	mean	L-	L+	mean	L-	L+	mean	L-	L+
CSBP	141	99	171	165	120	210	193	162	273
CDBP	87	69	102	102	86	118	113	97	129
CMSBP	175	137	219	175	123	227	204	154	279
Sokoloff	20.9	10	39	24	10	45	32	19	52
RA f(yes)	0			0			0.5		
complications f(yes)	0.03			0.08			0.17		
Indication for Treatment	category 1 (19 patients)			category 2 (19 patients)			category 3 (5 patients)		
CSBP	131.6	99	151	154.5	129	190	174	154	273
CDBP	82.6	69	90	93.5	81	102	106.2	95	121
risks	1.43	1	2.4	1.61	1	2.8	1.93	1	3
ant par f(yes)	0.74			0.64			0.59		
signs	1.31			1.21			1.7		

CDBP = corrected diastolic blood pressure
CSBP = corrected systolic blood pressure
CMSBP = corrected maximum systolic blood pressure
RA = repolarization abnormalities on EKG
complications = history of cardiac failure + history of coronary disease + arteritis of lower limbs + confirmed carotid stenosis + history of stroke
ant par = antecedents in parents (history of high blood pressure + history of stroke + history of coronary failure)
signs = headaches + sweating + palpitations + carotid murmur + lumbar murmur

10.1.6   Fuzzy Rule Construction and Prior Sensitivity Analysis

Fuzzy rules that reproduce the target, here the average coded answer, are sought. In the present case only the fuzzy "AND" appears to be needed. The rules may be constructed as follows. Let a patient i be represented by an experimental parameter value u_i for parameter n. This is then transformed into the membership function values $\mu_{n,k}^j(u_i)$, where n is the experimental parameter, and $k(k = 1, 2, 3$ or $k = -1, 0, 1)$ is a category index. Then the overall index for category k, patient i and question j is defined as

$$\text{overall category index} = V_i(j, k) = \prod_{n=1}^{N'} \mu_{n,k}^j(u_i) \qquad (10.1.6)$$

where N' is the number of selected parameters. The answer $A_i(j)$ is calculated as the defuzzified normed weighted sum combination

$$A_i(j) = \frac{\sum_{k=1}^{3} k V_i(j, k)}{\sum_{k=1}^{3} V_i(j, k)} \qquad (10.1.7)$$

A prior sensitivity analysis is then performed in order to determine the combination of parameters that minimizes the sum of squared errors $S^2(r)$ defined in Eq. 10.1.2. Table 10.3 shows the result of this procedure applied to the analysis of the two questions considered in our example. The three steps of the procedure are specified below:

Step 1. Apply the fuzzy rule, as defined by Eq. 10.1.6 and 10.1.7 to the whole set of selected parameters listed in Table 10.1.

Step 2. Remove successively one parameter at a time, and observe the effect of this removal on $S^2(j, r)$. The results are presented for question $j = 1$ and fuzzy rule $r = 0$ in Table 10.1.5 in the order of decreasing discriminant power of the parameter. For example, removing parameters CDBP and CSBP substantially increases $S^2(1, 0)$; RA and Sokoloff seems to have no influence at all; whereas removing the last two parameters, complications and MCSBP, decreases $S^2(1, 0)$.

Step 3. Find the parameter combination that minimizes $S^2(j, 0)$. For this purpose Eq. 10.1.2 is applied to increasing subsets of parameters taken in the order established in Step 2. In the present case, the optimal subset of parameters is CSBP + CDBP, as shown in Table 10.3.

Table 10.3. Example of prior sensitivity analysis: severity (r=0)

	$S^2(1,0)$
Step 1: calculation of $S^2(1,0)$ (Eq.10.1.2) for all selected parameters	
6 parameters:	
CDBP, CSBP, CMSBP, Sokoloff, RA, complications	4.2
Step 2: calculation of $S^2(1,0)$ for 5 out of 6 selected parameters.	
The results are presented in decreasing $S^2(1,0)$ order.	
Note that removing the last two	
parameters diminishes $S^2(1,0)$.	
They will not be retained	
for further analysis.	
5 parameters (without CDBP)	8.3
5 parameters (without CSBP)	6.0
5 parameters (without RA)	4.3
5 parameters (without Sokoloff)	4.3
5 parameters (without complications)	4
5 parameters (without CMSBP)	3.7
Step 3: constructing the final parameter set,	
by calculating $S^2(1,0)$	
for an increasing number of parameters	
taken in the order established	
in the second step.	
The minimum value, 3.2, is obtained	
with the first two parameters.	
Note hat adding the third	
one does not improve the score,	
and adding the fourth one worsens it.	
CDBP	5.1
CDBP+CSBP	3.2
CDBP+CSBP+complications	3.2
CDBP+CSBP+complications+Sokoloff	3.5

10.1.7 Application of the Fuzzy Rule to the Validation Set

Eq. 10.1.6 is now applied to the 50 patient files in the validation set. The results are compared to the average coded answers and to the individual experts' coded answers. The criteria $S^2(j,2)$ (Eq.10.1.2) and $D(r)$ (Eq.10.1.3) are calculated, with $r = 0$ for the fuzzy result and $r = 1, \ldots, 5$ for experts' answers. The results are illustrated in Fig. 10.3. The fuzzy rule approach appears to be able to reach the target very closely: on the average the fuzzy rule-based approach appears to yield answers with relatively small bias $D(j,r)$ and dispersion $S^2(j,r)$, as compared to the corresponding values of these two criteria for $j = 1,2$ and $r = 1, \ldots, 5$.

10.1.8 Discussion and Conclusions

This example of modeling the extremely difficult and complex problem of eliciting, encoding and reproducing multiple experts opinions (Heidel et al., 1992) demonstrates the usefulness of a fuzzy rule-based approach. The methodology is clearly not limited to the present case and can readily be extended to other cases of medical categorization, performed either by single or multiple experts. A sensitivity analysis has shown that the procedure is robust with respect to the shape of membership function. Besides, it can be used with dependent parameters, which is an important advantage in medical applications.

The high dispersion of experts' answers to two out of five questions from the primary set (Fig. 10.1) does not necessarily reflect an imprecision of experts' opinion, but may be due to different interpretations or attitudes. Actually this work was stimulated by the experts' desire to create a common classification system and facilitate the inter-expert communication.

Using the average answer as a target is a natural consequence of the decision to consider all experts' opinions as being of equal value. Otherwise, as stated earlier, a weighted average could be used or even, at the limit, the opinion of a single expert. The mean answer may appear as a weak target, but we must not forget the patient's point of view: if one physician recommends the antihypertensive treatment and another one advises against it, the patient is likely to say "I do not know what to do".

The membership functions have been defined on an experimental basis. A triangular form with peak at the mean value has been selected for sake of simplicity. The limited size of the calibration set (50 patient files) would not justify using more sophisticated forms as done, for example in Bárdossy et al. (1990) or as suggested in Section 2.3. In some cases a category was defined by so few points that the experimental information had to be completed by a subjective extension of the triangular membership function width. Given more data, a statistical procedure might be used to refine the definition of membership function (Civanlar and Trussel, 1986). Similarly, the frequency scheme (Eq. 10.1.6) used to incorporate the binary parameters may be refined for example by using TFN's.

The "AND" fuzzy rule used in this example is used with the product inference. The responses are combined using the normed weighted sum combination method and defuzzified using the fuzzy mean.

The fuzzy rule-based approach should not be applied automatically, without prior analysis. Out of five questions originally prepared, such a prior analysis shows that only three (hypertension severity, antihypertensive treatment indication, and estimation of coronary risk) appear to be suited for analysis using a fuzzy rule-based approach. The other two questions would require reconsideration by the experts.

In conclusion,

- the opinions of multiple medical experts have been assessed and combined using a fuzzy rule-based model

- a prior analysis has reduced the number of premises (experimental parameters) to very few for each question

- only one fuzzy rule and the small number of premises found in the prior analysis are needed to reach the target, i.e., the average coded experts answers, with relatively small dispersion and bias

- the patient state is described in a continuous way by an overall category index, although the initial categories are discrete

- the membership functions have been derived directly from experimental data

- the procedure may easily be extended to other cases of medical categorization.

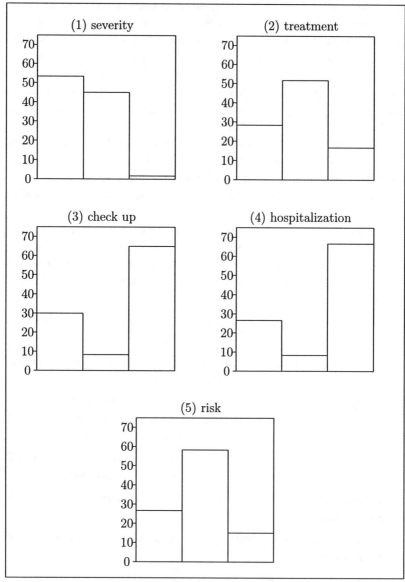

Figure 10.1. Histograms of maximum differences between the
five experts' coded responses to questions 1,...,5: (1) severity
of hypertension, (2) indication of antihypertensive treatment,
(3) indication for high blood pressure etiological check-up, (4)
indication for hospitalization, (5) estimation of coronary risk

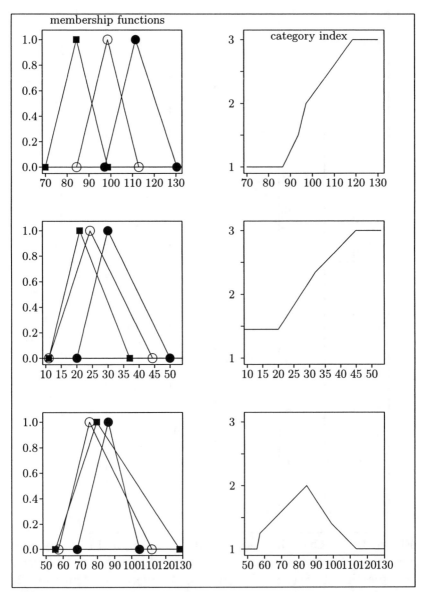

Figure 10.2.   Examples of discriminant power analysis ex-
tracted from the first question data set (severity). The left side
column shows examples of membership functions for three parameters, the right side column the corresponding partial category index as given by Eq. 10.4.

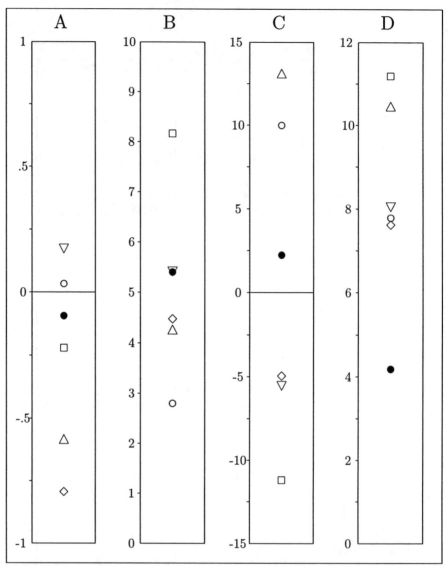

Figure 10.3. Criterion $D(j,r)$ and $S^2(j,r)$ applied to fuzzy rule based results and to experts' answers: **A:** Question 1, $D(j,r)$; **B:** Question 1, $S^2(j,r)$; **C:** Question 2, $D(j,r)$; **D:** Question 2, $S^2(j,r)$;

$\bullet$ = fuzzy
$\square$ = expert 1
$\triangle$ = expert 2
$\diamond$ = expert 3
$\triangledown$ = expert 4
$\circ$ = expert 5

10.2 Sustainable Reservoir Operation

The purpose of this section is to construct a sustainable reservoir operation rule by using a fuzzy rule-based model, which is simple, accurate, and representative of the system and is illustrated by a case study. A detailed description of the application can be found in Shrestha et al. (1995). Recent environmental concerns have focused on present as well as future environmental degradation, which is a consequence of shortsightness in present decision-making processes. These concerns have given rise to the new concept of sustainability. It has been agreed by now (at least in principle) that large systems, especially the ones that exploit natural resources, should be sustainable. The planning and operation of these systems should not only consider the benefits towards present day economics but also should prevent the adverse environmental effects on both present and future conditions. More of the philosophical backgrounds on the sustainable development can be found in Bruce 1992; Hatch 1992; Haimes 1992; and Plate 1993.

A real-world reservoir operation model can be very complex. It has to incorporate all the imprecisions in its inputs, and the output should fulfill all the constraints of the system. It especially should meet various demands without violating the physical constraints of the system. A fuzzy rule-based modeling for a reservoir operation is a simple approach that operates on an "IF - THEN" principle, where the "IF" is a vector of fuzzy premises such as the present state of reservoir pool elevation, the incoming inflow, the forecasted demand and the time of the year. The "THEN" is a fuzzy consequence such as actual release from the dam. The construction of fuzzy rules also incorporates the experience of the reservoir operator (expert of the system).

10.2.1 Sustainable Development

In the present study, we are seeking sustainable reservoir operation, namely sustainable operation rules that should always or nearly always fulfill reservoir purposes. The rules must be flexible so that they can be upgraded to reflect both societal (demand) and climatological (supply) changes. Let us look at an example given in Plate (1993). We know that the hydrological cycle is a random process; therefore, times of drought are unavoidable and a system that cannot always deliver water when needed is a possible outcome; its probability of

occurrence can be assessed by calculations based on simulation runs
with long time-series of inflows. Supply shortages should be met without undue impact on society. The rule therefore should be such that
it restricts the use of water early enough to be effective during the
forecasted water shortages. The active storage capacity of a reservoir
should not be reduced at all.

Five operating principles of sustainable development are described
in Haimes (1992) as follows:

1. Multiobjective analysis: sustainable development requires both
   economic development and environmental protection objectives
   and must achieve an acceptable balance between them.

2. Risk analysis, including risk of extreme events: the environmental
   risk associated with economic development must be assessed and
   special consideration must be given to extreme events that occur
   with very low probabilities but with dire and possible catastrophic
   consequences.

3. Impact analysis: the impacts of current decisions on future op-
   tions must be evaluated by strategic planning and/or sensitivity
   analysis.

4. Consideration of multiple decision makers and constituencies (e.g.,
   regions, sectors, socioeconomic and political subdivisions): this is
   critical from two perspectives — comprehensiveness and practi-
   cability/operationality.

5. Accounting for interaction among a system's components and between the system and its environment: theoretical-philosophical
   and operational-pragmatic dimensions are both important and
   consequential.

In the present study, the sustainability concept is considered a function reliability, resilience and vulnerability, as defined in Shrestha et
al. (1995). There another important figure of merit is defined, namely
engineering risk, which is a function of reliability, incident period and
repairability performance indices.

10.2.2   Fuzzy Reservoir Model and Results

In modeling reservoir operation by a fuzzy rule-based scheme, the
premises are: present state reservoir pool elevation (or total storage
of the reservoir), the incoming flow (the net flow into the reservoir is
inflow minus losses, especially evaporation and seepage), forecasted
demand states and time of the year. The consequence is the actual

release to meet the demands. The premises pool elevation and net inflows and the consequence, i.e., the actual power release are considered to be TFNs. Therefore, the rule-system structure is given as follows:

If pool elevation is $A_{i,1}$ ⊙ forecasted net inflow is $A_{i,2}$

⊙ forecasted demand state is $A_{i,3}$⊙ time of the year is $A_{i,4}$

then actual power release is B_i $\qquad$ (10.2.1)

here ⊙ denotes an "AND" combination. Split sampling is used to calibrate and validate the fuzzy rules. Calibration is done by using a training set and counting algorithm as described in Section 10.2.3. The system or mass balance equation and the physical or boundary conditions for reservoir operation are given as follows:

System or Mass Balance Equation:

$$S_{t+1} = S_t + I_t - R_t - L_t \qquad (10.2.2)$$

$$\begin{aligned} S_t, \ I_t, \ R_t \ &= \ \text{storage, inflow and release at time t} \\ L_t \ &= \ \text{evaporation and seepage losses at t} \end{aligned}$$

Physical or Boundary Conditions:

$$\begin{aligned} S_{t,\min} &\leq S_t \leq S_{t,\max} \\ R_{t,\min} &\leq R_t \leq R_{t,\max} \end{aligned} \qquad (10.2.3)$$

$$\begin{aligned} S_{t,\min} \ &= \ \text{min. storage level at time t} \\ S_{t,\max} \ &= \ \text{max. storage level at time t} \\ R_{t,\min} \ &= \ \text{min. release at time t} \\ R_{t,\max} \ &= \ \text{max. release at time t (capacity of d/s channel)} \end{aligned}$$

A fuzzy rule system is constructed for Tenkiller Lake dam in the Illinois river basin (Oklahoma). The dam is located at river mile 12.8 on the Illinois river about 7 miles northeast of Gore, and about 25 miles south of Tahlequah, as shown in Fig. 10.4. The reservoir is built to serve the two main purposes of flood control and hydropower generation. The project also provides water supply, recreation facilities and fish and wildlife conservation.

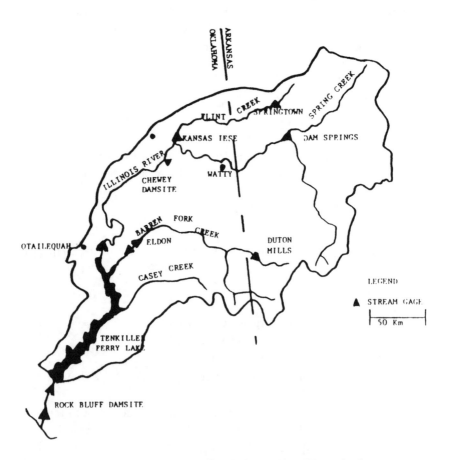

Figure 10.4.   Tenkiller Lake and its Watershed

The top of Tenkiller lake dam is at elevation 677.2 which is about 197 feet above the streambed. The top of flood control pool is at elevation 667.0, the top of conservation pool is at elevation 632.0, the top of inactive pool is at elevation 594.5. The lake area at the top of conservation pool is about 12,900 acres. The total conservation storage is about 371,000 acre feet. The reservoir stretches about 34 miles with 130 miles of shoreline. The other pertinent data related to the lake can be found in U.S. Army Corps of Engineers (1982). Hydrological data (pool elevations and volume, inflow, rainfall, evaporation loss) and demand data (power release and total release) have been obtained from U.S. Army Corps of Engineers, IWR, Fort Belvoir, Virginia. The data set contains the daily values from year 1980 to 1992.

Year 1989 is used for validation of fuzzy rules and rest of the data is used for training the rules (calibration). As explained in the previous section, the weighted counting algorithm is selected to generate the rules. Nine TFNs are used to describe pool elevations from 620 to 677.2 NG-ft., and eight TFNs are defined to cover inflow ranging from 0 to 180,000 DSF (357,000 CFS). Supports of the TFNs for both pool elevations and inflows are chosen such that these TFNs cover the entire range of the values of pool elevations and inflows and the two neighboring TFNs have at least 25% overlapping, as suggested in Kosko (1992). The rules are trained for each month separately along with other premises the forecasted demand state. Five power demand states: Low, Medium-Low, Medium, Medium-High, and High states are used. Instead of using the actual forecast demand, only the information regarding the power states using threshold values are used. Both premises: the time of year and forecasted demand states play important roles while calibrating the fuzzy rules. Note that since product inference is used for DOFs calculation, care must be taken in choosing ϵ value. Higher ϵ value will result in an incomplete system (a system without a response). This situation can be avoided by either using a larger data set (which is not possible given the data on hand) or by using the wider fuzzy numbers, i.e., fuzzy numbers with larger supports. The problem in using wider TFNs is that they yield larger errors in the response. The tradeoff should be such that the FRB describe the physical process as well as possible without losing the completeness of the rule system. Note that if we use maximum or additive inference instead of product inference used in this study, the rule system will always be complete (Proposition 4.2). As mentioned earlier, the selection of DOF inference should be entirely based on the physical phenomenon of a system. Note that we used an additive combination for response and provided we have at least a single valid response, we will have the response rule system always complete.

Fig. 10.5 shows the validation of fuzzy rules, that is, the observed vs. fuzzy rule-based power releases in the year 1989. The actual numerical values for January 1989 are given in Table 10.1 as an example of the model input and output. The figure shows that the fuzzy rules are robust and do not change much for small changes in the premise values. FRB releases match the actual reservoir releases at almost all times of the year, especially for the last four months: namely from September to December 1989 when the FRB generated releases are as low as the actual releases.

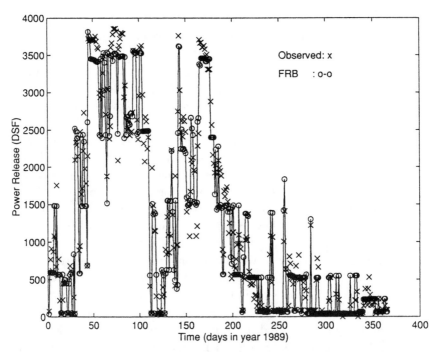

Figure 10.5. Comparison between FRB based releases and actual release

10.2.3 Discussion and Conclusions

Reservoir release from any real-world project can be very complex. Besides the physical constraints (such as dam, spillway and downstream channel capacity), the demands constraints to be met are also quite important. Further, imprecisions in forecasted inflow and

Table 10.4. FRB validation for January 1989

Days	Pool Elevation (NG-ft.)	Inflow (DSF)	FRB Release (DSF)	Actual Release (DSF)
1	631.56	1000	39.2	74.0
2	631.68	1000	593.2	582.0
3	631.75	1000	593.3	768.0
4	631.76	1000	593.4	911.0
5	631.74	1000	593.3	617.0
6	631.78	1000	593.4	875.0
7	631.79	500	1480.3	1078.0
8	631.73	500	1479.3	1050.0
9	631.65	500	567.5	670.0
10	631.63	500	1477.6	1754.0
11	631.42	500	565.8	549.0
12	631.42	500	565.8	771.0
13	631.42	500	565.8	219.0
14	631.45	500	53.5	42.0
15	631.51	500	53.4	41.0
16	631.54	500	566.7	227.0
17	631.57	300	444.8	445.0
18	631.66	400	495.7	445.0
19	631.54	400	495.8	450.0
20	631.52	200	448.1	468.0
21	631.51	300	54.2	45.0
22	631.54	300	54.2	45.0
23	631.54	300	445.1	530.0
24	631.51	500	566.5	683.0
25	631.47	1200	593.3	620.0
26	631.57	700	52.3	42.0
27	631.69	1000	39.2	38.0
28	631.86	2500	838.3	260.0
29	632.25	2600	43.0	45.0
30	632.68	4000	2517.5	2095.0
31	632.94	3200	2483.6	2331.0

evaporation losses also play important roles. Most important, the experience of the reservoir operator also has a large role in dam releases. The FRB model tries to capture all these in its premises and rules output response tries to imitate the actual response. The construction of the fuzzy rules uses a training set with a simple algorithm. We do not require use of a complex model to get similar results. Some advantages of the FRB model are:

1. The FRB model is very simple. Fuzzy rule construction (or calibration) requires a training set. The training set can be either a set of observed data (as in this study) or a set of outputs from a physical model as in Bárdossy and Disse (1993).

2. The model is transparent and easy to understand due to its rule-based structure, which mimics the human way of thinking.

3. The computation of response by fuzzy rules is easy and requires very little time.

4. The model is robust: the response does not change much with small change in the premise values.

References

Bahill, A.T. (1991): *Verifying and validating personal computer-based expert systems,* Prentice-Hall, Englewood Cliffs, NJ.

Bárdossy, A. (1990): Notes on fuzzy regression - *Fuzzy Sets and Systems* **31/1**: 65–75.

Bárdossy, A. and E. J. Plate (1992): Space-time model for daily rainfall using atmospheric circulation patterns - *Water Resources Research,* **28**: 1247–1259.

Bárdossy, A. and H.J. Caspary (1990): Detection of climate change in Europe by analyzing European atmospheric circulation patterns from 1881 to 1988. *Theoretical and Applied Climatology* **42**: 155–167.

Bárdossy, A., H. Muster, L. Duckstein and I. Bogárdi (1993c): Knowledge based classification of circulation patterns for stochastic precipitation modeling, *Proceedings International Conference on Stochastic Methods in Hydrology and Environmental Engineering,* Waterloo, Ontario, June, 201–215.

Bárdossy, A. and I. Bogárdi (1989): Fuzzy fatigue life prediction, *Structural Safety* **6**: 25–38.

Bárdossy, A., I. Bogárdi, and L. Duckstein (1991): Fuzzy regression in hydrology, *Water Resources Research* **26/7**: 1497–1508.

Bárdossy,A., I. Bogárdi, L. Duckstein (1993b): Fuzzy nonlinear regression analysis of dose - response relationships, *European Journal of Operations Research,* **66**, 36–51.

Bárdossy A., I. Bogárdi, and W. E. Kelly, (1990): Kriging with imprecise (fuzzy) variogram, Part I: Theory, Part II: Application - *Mathematical Geology* **22/1**: 63–94.

Bárdossy, A and L. Duckstein (1992): Analysis of a karstic aquifer management problem by fuzzy composite programming, *Water Resources Bulletin* **28/1**: 63–74.

Bárdossy, A., L. Duckstein and I. Bogárdi (1992b): Fuzzy classification of atmospheric circulation patterns, *SIE working paper.*

216

Bárdossy, A., L. Duckstein and I. Bogárdi (1993a): Combination of fuzzy numbers reflecting expert opinions, *Fuzzy Sets and Systems* **57**: 173–181.

Bárdossy, A., L. Duckstein and I. Bogárdi (1994): Fuzzy rule-based classification of atmospheric circulation patterns. *int. j. climatology,* in press.

Bárdossy A. and M. Disse, (1993): Fuzzy rule-based models for infiltration, *Water Resources Research,* **29**, 373–382.

Bárdossy, A., R. Hagaman, I. Bogárdi, L. Duckstein (1992a): Least squares regression: theory and application, in *Fuzzy regression analysis,* , J. Kacprzyk and M. Fedrizzi (eds.), Omnitech Press, Warsaw, 181–193.

Baur, F., P. Hess, und Nagel, H. (1944): *Kalender der Großwetterlagen Europas 1881-1939.* Bad Homburg.

Berger, J. O. (1985): *Statistical decision theory and Bayesian analysis,* Springer Verlag, N.Y. 617p.

Bernier, J., (1994): Statistical detection of change - In: *Natural Resources Management,* L. Duckstein and E. Parent (eds). Dordrecht, The Netherlands.

Bernier, J., (1994): Quantitative analysis of uncertainties in water resource: for predicting the effects of changes, In: *Natural Resources Management,* L. Duckstein and E. Parent (eds.), Dordrecht, The Netherlands.

Bezdek, (1981): *Pattern Recognition with Fuzzy Objective Function Algorithms,* Plenum Press, New York, 256p.

Blinowska, A. and L. Duckstein, (1993): Medical applications of fuzzy logic-fuzzy patient state in arterial hypertension analysis- *IEEE EMBS Trans.* (to appear).

Blinowska, A., L. Duckstein and J. Verroust, (1992): Patient state categorization by a distance measure between fuzzy numbers applied to peripheral neuropathy, *Proc. XIVth An. IEEE EMBS Conf.,* Paris, France.

Bogárdi, I., I. Matyasovszky, A. Bárdossy and L. Duckstein,(1994): Estimation of local climate factors under climate change, In: *Natural Resources Management,* L. Duckstein and E. Parent (eds), Dordrecht, The Netherlands.

Bradley, R.S., R.G. Barry, and G. Kiladis (1982): Climatic fluctuations of the Western United States during the period of instrumental records, *Final Report to the National Science Foundation,* University of Massachusetts, Amherst.

Bruce, J. P. (1992): Meteorology and Hydrology for Sustainable Development, *World Meteorological Organization No. 769,* Secretariat of the WMO, Geneva, Switzerland.

Casti, J. (1990): *Searching for certainty,* Morrow, N.Y.

Casti, J. (1993): *Reality rules I, II,* J. Wiley & Sons, NY.

Casti, J., J. Kempf, L. Duckstein, and M. Fogel (1979): Lake ecosystems: A polyhedral dynamics representation, *J. Ecological Modeling,* **7**: 223–237.

Civanlar, M. R. and H. J. Trussel (1986): Constructing membership functions using statistical data, *Fuzzy Sets and Systems,***18**: 1–13.

Diamond, Ph. (1988): Fuzzy least squares, *Information Sciences,* **46**: 141–157.

Draper, N. R. and H. Smith, (1990): *Apllied Regression Analysis.* 10th edition Wiley N. Y.709 p.

Dubois, D. and H. Prade (1980a): *Fuzzy sets and systems: Theory and applications,* New York, Academic Press, 393p.

Dubois, D. and H. Prade (1980b): New results about properties and semantics of fuzzy set operators, In *Fuzzy Sets* , (P.P. Wang and S.K. Chang, Eds.), New York, Plenum Press, 59–75.

Dubois, D. and H. Prade (1986): Fuzzy sets and statistical data - *European Journal of Operational Research,* **25**: 345–356.

Dubois, D., and H. Prade (1988): *Possibility Theory: An Approach to Computerized Processing of Uncertainty* New York, Plenum Press, p. 263.

Dubois, D. and H. Prade (1991): Fuzzy sets in approximate reasoning, Part 1: Inference with possibility distributions, *Fuzzy Sets and Systems* **57**: 173–181.

Dubois, D. and H. Prade (1993): Fuzzy sets and probability: Misunderstandings, bridges and gaps, in *Proceedings 2nd IEEE Int. Conf. Fuzzy Systems (FUZZ-IEEE'93),* San Francisco, CA

Dubois, D. and H. Prade (1994a): Fuzzy sets - a convenient fiction for modeling vagueness and possibility, *IEEE Transactions on Fuzzy Systems,* **2**: 16–21.

Dubois, D. and H. Prade (1994b): Basic issues on fuzzy rules and their application to fuzzy control In *Fuzzy Logic and Fuzzy control* (Edited by D. Driankov, P.W. Eklund and A. Ralescu) Lecture Notes in Artificial Intelligence vol. 833 Springer Verlag, 3–13.

Dubois, D., M Grabisch and H. Prade (1995): Gradual rules and the approximation of functions *Proc. 2nd International Conf. On Fuzzy Logic and Neural Networks, Iizuka, Japan, July 1992, 629-632 (Extended version Gradual rules and the approximation of control laws* to appear in Theoretical aspects of Fuzzy Control (Edited by H. Nguyen and M. Sugeno), Wiley.

Duckstein, L. (1994): Engineering risk under non-steady conditions: Bayes and fuzzy logic approaches, *Keynote lecture UNESCO-IHP Conference,* Karlsruhe, Germany, 28-30, June

Duckstein, L., A. Blinowska, and J. Verroust (1993): Fuzzy classification of patient state with application to electrodiagnosis of peripheral polyneuropathy, *IEEE EMBS Trans.* (to appear).

Duckstein, L., A. Blinowska, J. Verroust, and P. Degoulet (1993): Fuzzy modeling of patient state: distance-based versus rule-based approach - *Proceedings IEEE-FUZZ 93,* San Francisco.

Duckstein, L. and E. Parent (1994): Systems engineering of natural resources under changing physical conditions: a framework for reliability and risk, In: *Natural Resources Management* L. Duckstein and E. Parent (eds). Kluwer, Dordreche, The Netherlands.

Duckstein, L, I. Bogárdi and A. Bárdossy (1989): Fuzzy set membership, prior probability and value function, *SIE working paper.*

Duckstein, L., and K. Heidel (1988): Estimation of fuzzy set membership functions using value function transformation, *Fourth International Conference on the Foundations and Applications of Utility, Risk and Decision Theory,* Hungary.

Ge, S. and W. Laurig (1993): Properties of membership functions for the perception of forces with respect to psychophysiological aspects, in *Proceedings of the First European Congress on Fuzzy and intelligent Technologies.* Aachen September 7-10, 1993, pp. 657-662.

Green, W.H. and Ampt, G.A. (1911): Studies of soil physics, 1. The flow of air and water through soils, *J. Agric. Sci.,* **4,** 1-24.

Haimes, Y. Y. (1992): Sustainable Development: A Holistic Approach to Natural Resource Management, *IEEE Transactions on Systems, Man, and Cybernetics,* Vol. 22, No. 3, pp. 413-417, May/June.

Hatch, H. J. (1992): Accepting the Challenge of Sustainable Development, *The Bridge - Nat. Acad. Eng.,* Vol. 22, No. 1, Spring.

Plate, E. J. (1993): Sustainable Development of Water Resources: A Challenge to Science and Engineering, *Water International,* , **18:** 84–94.

Kosko, B. (1992): *Neural Networks and Fuzzy Systems: A Dynamical Systems Approach to Machine Intelligence,* Prentice Hall, NJ, 449p.

Heidel, K., R. Ferrell and L. Duckstein (1992): Development of group consensus indices using fuzzy numbers, *VIth Intern. Conf. on the Foundations and Applic. of Utility, Risk and Decision Theory,* Canchan, France.

Hess, P. und Brezowsky, H. (1969): *Katalog der Großwetterlagen Europas. Berichte des Deutschen Wetterdienstes* **113** (15), 2. neu bearbeitete und ergänzte Aufl., Offenbach a. Main, Selbstverlag des Deutschen Wetterdienstes.

Hisdal, E. (1994): Interpretative versus perspective fuzzy theory, *IEEE Transactions on Fuzzy Systems,* **2**: 22–26.

Holmblad, L. P. and J. J. Ostergaard (1982): Control of a cement kiln by fuzzy logic, In: *Fuzzy Information and Decision Processes,* M. M. Gupta and E. Sanchez (eds), North Holland, Amsterdam.

Johnson, S. C. (1967): Hierarchical clustering schemes, *Psychometrika* **32**: 261–274.

Kaufmann, A. and M. M. Gupta (1991): *Introduction to fuzzy arithmetic: Theory and applications,* Van Nostrand Rheinhold, New York.

Klir, G. J., and T. A. Folger (1988): *Fuzzy sets, uncertainty and information,* Prentice Hall, New Jersey.

Klir,G. J. (1994): On the alleged superiority of probabilistic representation of uncertainty - *Fuzzy Sets and Systems* **2/1**: 27–31.

Kosko, B. (1992): *Neural networks and fuzzy systems: a dynamical systems approach to machine intelligence,* Prentice Hall, NJ, 449p.

Kosko, B. (1994): The probability monopoly — *Fuzzy Sets and Systems* **2/1**: 32–33.

Krick, J.P. (1944): Synoptic weather types of North America, *California Institute of Techn.,* Pasadena, California, 237p.

MacQueen, J. (1967): Some methods for classification and analysis of multivariate observations, *Fifth Berkeley Symposium on Mathematics,* **1**: 281–298.

Mamdani E.H. (1977): Application of fuzzy logic to approximate reasoning using linguistic systems, *IEEE Transactions on Comput.,* **26**: 1182–1191.

Mamdani,E. H., J. J. Ostergaard and E. Lembessis. (1983): Use of fuzzy logic for implementing rule-based control of industrial processes. In: *Advances in Fuzzy Sets,Possibility Theory, and Applications.* Plenum Press, 307-323.

Muster,H., A. Bardossy and L. Duckstein (1994): Adaptive neuro-fuzzy modeling of a non-stationary hydrologic variable. *Proceedings,* Int. Sympos. on Water Resources in a Changing World, Karlsruhe, Germanu, II-221 to II-230.

Matyasovszky, I., I. Bogárdi, A. Bárdossy, and L. Duckstein (1992): Hydroclimatological modeling of drought under climate change, *Proceedings, ICID 16th European Regional Conference Budapest,* June.

Ozelkan, E, F. Ni and L. Duckstein (1994): Fuzzy rule-based approach for analyzing the relationship between monthly atmospheric circulation patterns and extreme precipitation, *working paper, SIE, univ. of AZ, Tucson.*

Parent, E., and L. Duckstein (1994): Reliability and risk in the engineering of natural resources under changing physical conditions: state of the art, In: *Natural Resources Management,* L. Duckstein and E. Parent (eds), Dordrecht, The Netherlands.

Pesti, G., B. P. Shrestha, L. Duckstein, and I. Bogárdi (1994) Estimation of the impacts of global climate change on regional droughts in a fuzzy logic framework, *International Specialty Conference on Global Climate Change: Science, Policy and Mitigation Strategies,* Phoenix, AZ, April.

Plate, E. J. (1992): Sustainable Development of Water Resources: A Challenge to Science and Engineering, *Proceedings,* Int. Symposium of the IAHR, Kyoto, Japan.

Richards, L.A. (1931): Capillary conduction of liquids through porous media, *Physics,* **1**, 318–333

Richardson C.W. (1981): Stochastic simulation of daily precipitation, temperature and solar radiation, *Water Resources Research,* **17**, 182–190.

Rommelfanger, H. (1993): Fuzzy-Logik *OR-Spektrum,* **15**, 31–41.

Schmucker, K.J. (1984): *Fuzzy sets, natural language computations, and risk analysis.* , Computer Science Press, Rockville, 192p.

Schweizer, B. and A. Sklar (1961): Associative functions and statistical triangle inequalities, *Publ. Math. Debrecen,* **8**, 169–186.

Shingu, T. and E.Nishimori, (1989): Fuzzy based automatic focusing system for a compact camera, In: *Proceedings of Third IFSA Congress,*436–439.

Shrestha, B. P. and L. Duckstein (1993): Fuzzy reliability measures, *SIE Working paper.*

Shrestha, B. P., L. Duckstein, and E. Z. Stakhiv (1995): Fuzzy rule-based modeling of sustainable reservoir operation, *IAHS Boulder General Meeting,* July.

Stevens, S.S. (1975): *Psychophysics,* John Wiley & Sons, New York.

Sugeno, M. (editor) (1985): *Industrial applications of fuzzy control* North-Holland, Amsterdam.

Sugeno, M. and M. Nishida (1985): Fuzzy control of model car *Fuzzy Sets and Systems* **16**: 103–113.

Sugeno, M. and T. Takagi (1983): Multi-dimensional fuzzy reasoning, *Fuzzy Sets and Systems* **9**: 313–325.

Sugeno, M. and T. Yasukawa (1993): A Fuzzy-Logic-Based approach to qualitative modeling, *IEEE Transactions on Fuzzy Systems,* **1/1**: 7–31.

Tagaki, M. (1990): Fusion technology of fuzzy theory and neural networks - survey and future directions, *Proceedings First International Conference on Fuzzy Logic & Neural Networks (IIZU KA "90),* 13–26.

Tagaki, H. and I. Hayashi (1990): NN-driven fuzzy reasoning, *Int'l J. Approximate Reasoning,* **5/3**: 191–212.

Türksen, I.B. (1991): Measurement of membership functions and their acquisition, *Fuzzy Set and Systems,* **40/1**: 5–34.

US Army Corps of Engineers, Southwestern Division, Tulsa District (1982): Restudy of Tenkiller Ferry Lake Illinois River, Oklahoma, *Draft Survey Report and Environmental Impact Statement,* .

Van Genuchten, M. Th. (1980): A closed-form equation for predicting the hydraulic conductivity of unsaturated soils, *Soil Sci. Soc. Am.,* 44: 892–898.

Wang, F. Y. and D. D. Chen (1993): Rule generation and modification for intelligent controls using fuzzy logic and neural networks, *SIE Working paper.*

Wang, L. X. and J. M. Mendel, (1990): Generating fuzzy rules from numerical data with applications. *Tech. Rep. USC-SIPI 169,* univ of so. CA, Los Angeles.

Waterstone, M. (1994): Institutional analysis and water resources management, In: *Natural Resources Management,* L. Duckstein and E. Parent (eds.) Kluwer, Dordrecht, The Netherlands.

Wymore, A. W. (1993): *Model Based Systems Engineering,* CRC Press, Boca Raton, FL, 736p.

Yager,R. R. , (1980): On a general class of fuzzy connectives. *Fuzzy sets and systems,* 4: 235-242.

Yager, R. R. (1991): Connectives and quantifiers in fuzzy sets, *Fuzzy Sets and Systems,* **40**, 35–75.

Yager, R. R. and D. P. Filev (1992): SLIDE: A simple adaptive defuzzification method, *IEEE Transactions on Fuzzy Systems,* **1**: 69–78.

Yamakawa, T. (1989): Stabilization of an inverted pendulum by a high speed fuzzy controller hardware system, *Fuzzy Sets and Systems,* **32**, 161–180.

Yarnal, B. (1984): A procedure for the classification of synoptic weather maps from gridded atmospheric pressure surface data. *Computers and Geosciences* **10**: 397–410.

Yarnal, B. (1993): *Synoptic climatology in environmental analysis: Studies in climatology series,* Edit. by Gregory, S., Belhaven Press, London and CRC Press, Boca Raton, Florida.

Yasunobu, S. and S. Miyamoto (1985): Automatic train operation system by predictive fuzzy control, In: *Industrial Applications of Fuzzy Control,* M. Sugeno (ed). North Holland, Amsterdam, 1–18.

Zadeh, L. A. (1965): Fuzzy sets, *Information and Control,* **8**: 338–353.

Zadeh, L. A. (1972): A fuzzy set-theoretic interpretation of linguistic hedges, *Journal of Cybernetics,* **2**, 4–34.

Zadeh, L. A. (1973): Outline of a new approach to the analysis of complex systems and decision processes, In: *IEEE Trns.* **SMC 3**: 28–44.

Zadeh, L. A. (1975): The concept of a linguistic variable and its application to approximate reasoning - I., *Information Sciences,* **8**: 199–249.

Zadeh, L. A. (1978): Fuzzy sets as a basis for a theory of possibility, *Fuzzy Sets and Systems,* **1**, 3–28.

Zadeh, L. A. (1983): A computational approach to fuzzy quantifiers in natural languages, *Comput. Math. Appl.,* **9**, 149–184.

Zhang, Xiaohui, and L. Duckstein (1993): Combination of radiologists diagnosis by a fuzzy number approach, *Proceedings IEEE - FUZZ 93,* San Francisco, IEEE- Neural Networks Council, 1293–1298.

Zimmermann, H. J. (1985): *Fuzzy set theory and its application,* Kluwer Nijhoff Publishing, Dordrecht.

Proofs of Selected Propositions

Proposition A.1 *Consider a fuzzy set A with a piecewise linear membership function having breakpoints: $x_0 < x_1 < \ldots < x_L$. The fuzzy mean of A can then be calculated as:*

$$M(A) = \frac{\sum_{l=0}^{L-1} (x_{l+1} - x_l) \left(\frac{2x_{l+1}+x_l}{6} \mu(x_{l+1}) + \frac{x_{l+1}+2x_l}{6} \mu(x_l) \right)}{\sum_{l=0}^{L} (x_{l+1} - x_l) \frac{\mu(x_{l+1})+\mu(x_l)}{2}}$$

Proof:
The integration for the calculation of the fuzzy mean can be performed for each interval separately:

$$M(A) = \frac{\int\limits_{-\infty}^{+\infty} t\mu_A(t)\, dt}{\int\limits_{-\infty}^{+\infty} \mu_A(t)\, dt} == \frac{\sum_{l=0}^{L-1} \int\limits_{x_l}^{x_{l+1}} t\mu_A(t)\, dt}{\sum_{l=0}^{L-1} \int\limits_{x_l}^{x_{l+1}} \mu_A(t)\, dt}$$

As the membership function $\mu_A(x)$ is linear in $[x_l, x_{l+1}]$ it can be written as:

$$\mu_A(x) = \frac{\mu_A(x_{l+1}) - \mu_A(x_l)}{x_{l+1} - x_l}(x - x_l) + \mu_A(x_l) \quad x \in [x_l, x_{l+1}]$$

Thus

$$\int\limits_{x_l}^{x_{l+1}} \mu_A(t)\, dt = \frac{\mu_A(x_{l+1}) + \mu_A(x_l)}{2}(x_{l+1} - x_l)$$

which explains the denominator.

$$\int\limits_{x_l}^{x_{l+1}} t\mu_A(t)\, dt = \int\limits_{x_l}^{x_{l+1}} \frac{\mu_A(x_{l+1}) - \mu_A(x_l)}{x_{l+1} - x_l}(t - x_l)t + \mu_A(x_l)t\, dt =$$

$$
= \frac{x_{l+1}^3 - x_l^3}{3} \left(\frac{\mu_A(x_{l+1}) - \mu_A(x_l)}{x_{l+1} - x_l} \right) -
$$

$$
- \frac{(x_{l+1}^2 - x_l^2)x_l}{2} \left(\frac{\mu_A(x_{l+1}) - \mu_A(x_l)}{x_{l+1} - x_l} \right) + \frac{x_{l+1}^2 - x_l^2}{2} \mu_A(x_l)
$$

after some calculations one has:

$$
\int_{x_l}^{x_{l+1}} t\mu_A(t)\, dt = (x_{l+1} - x_l) \left(\frac{2x_{l+1} + x_l}{6} \mu(x_{l+1}) + \frac{x_{l+1} + 2x_l}{6} \mu(x_l) \right)
$$

from which the formula follows immediately.

Proposition A.2 *The fuzzy mean $M(A)$ is continuous for fuzzy sets with the same bounded support $[-K, K]$: For each A and $\varepsilon > 0$ there is a $\delta > 0$ such that*

$$
|M(A) - M(B)| < \varepsilon
$$

if

$$
|\mu_A(x) - \mu_B(x)| < \delta
$$

for all $x \in [-K, K]$

Proof:
Using the finite support of the fuzzy sets A and B

$$
|M(A) - M(B)| = \left| \frac{\int_{-K}^{+K} t\mu_A(t)\, dt}{\int_{-K}^{+K} \mu_A(t)\, dt} - \frac{\int_{-K}^{+K} t\mu_B(t)\, dt}{\int_{-K}^{+K} \mu_B(t)\, dt} \right| =
$$

$$
= \left| \frac{\int_{-K}^{+K} t\, (\mu_A(t)M_A - \mu_B(t)M_B)\, dt}{M_A M_B} \right|
$$

where

$$M_A = \int_{-K}^{+K} \mu_A(t)\, dt$$

But as:

$$\left| \frac{\int_{-K}^{+K} t\left(\mu_A(t)M_A - \mu_B(t)M_B\right)\, dt}{M_A M_B} \right| =$$

$$= \left| \frac{\int_{-K}^{+K} t\left(\mu_A(t)M_A - \mu_A(t)M_B + \mu_A(t)M_B\mu_B(t)M_B\right)\, dt}{M_A M_B} \right| \leq$$

$$\leq \frac{\int_{-K}^{+K} |t|\left(\mu_A(t)2K\delta + \delta M_B\right)\, dt}{M_A M_B} \leq \frac{2K^2(M_A 2K\delta + \delta M_B)}{M_A M_B}$$

but as

$$M_A + 2K\delta \geq M_B \geq M_A - 2K\delta$$

one also has:

$$\frac{2K^2(M_A 2K\delta + \delta M_B)}{M_A M_B} \leq \frac{2K^2(M_A 2K\delta + \delta(M_A + 2K\delta))}{M_A(M_A - 2K\delta)}$$

This last quantity can be arbitrarily small if δ is small enough, which completes the proof.

Proposition A.3 *If the product inference rules (3.1.8-3.1.10) are used, and an additive combination method (3.8 or 3.9) is used then the rule*

If A_1 OR A_2 then B

can be replaced by three rules:

If A_1 AND A_2 then B

$$\text{If } A_1 \text{ AND } (NOT \ A_2) \text{ then } B$$

$$\text{If } (NOT \ A_1) \text{ AND } A_2 \text{ then } B$$

without a change in the result.

Proof:

The consequence for (a_1, a_2) is

$$(\mu_{A_1}(a_1) + \mu_{A_2}(a_2) - \mu_{A_1}(a_1)\mu_{A_2}(a_2)) \, \mu_B(x)$$

The consequence of the three rules is the fuzzy set

$$(\mu_{A_1}(a_1)\mu_{A_2}(a_2)) \, \mu_B(x) + (\mu_{A_1}(a_1)(1 - \mu_{A_2}(a_2))) \, \mu_B(x) +$$

$$+ ((1 - \mu_{A_1}(a_1))\mu_{A_2}(a_2)) \, \mu_B(x) =$$

$$= (\mu_{A_1}(a_1) + \mu_{A_2}(a_2) - \mu_{A_1}(a_1)\mu_{A_2}(a_2)) \, \mu_B(x)$$

Proposition A.4 *If the product inference rules (3.1.8-3.1.10) are used, and an additive combination method (3.8 or 3.9) is used then the rule*

$$\text{If } A_1 \text{ XOR } A_2 \text{ then } B$$

can be replaced by two rules:

$$\text{If } A_1 \text{ AND } (NOT \ A_2) \text{ then } B$$

$$\text{If } (NOT \ A_1) \text{ AND } A_2 \text{ then } B$$

without a change in the result.

Proof:

The consequence for (a_1, a_2) is

$$(\mu_{A_1}(a_1) + \mu_{A_2}(a_2) - 2\mu_{A_1}(a_1)\mu_{A_2}(a_2)) \, \mu_B(x)$$

The consequence of the two rules is the fuzzy set

$$(\mu_{A_1}(a_1)(1 - \mu_{A_2}(a_2)))\,\mu_B(x) + ((1 - \mu_{A_1}(a_1))\mu_{A_2}(a_2))\,\mu_B(x) =$$

$$= (\mu_{A_1}(a_1) + \mu_{A_2}(a_2) - 2\mu_{A_1}(a_1)\mu_{A_2}(a_2))\,\mu_B(x)$$

Proposition A.5 *Suppose the rules $s = 1, \ldots, S$ have positive fulfill-ment grades $\nu_s = D_s(a_1, \ldots, a_K) > 0$. Then for the combined answer B:*

$$\min_s \mathcal{D}_f(B_s) \leq \mathcal{D}_f(B) \leq \max_s \mathcal{D}_f(B_s)$$

Proof:
For the mean defuzzification and the additive decompositions the proof is evident, as the mean of B is a convex combination of the means B_i.

Proposition A.6 *A rule system $\mathcal{R}$ with fuzzy numbers as consequences and being used with the minimum combination is complete if and only if*

1. *for each $(a_1, \ldots, a_K) \in A$ there is a rule i such that $D_i(a_1, \ldots, a_K) > 0$ and*

2. *For any two rules i and j if there is an $(a_1, \ldots, a_K) \in A$ such that the degree of fulfillment $D_i(a_1, \ldots, a_K) > 0$ and $D_j(a_1, \ldots, a_K) > 0$ then $B_i \cap B_j \neq \emptyset$*

Proof:
Let $i_1, \ldots, i_J$ be the rules for which the degree of fulfillment $D_{i_j}(a_1, \ldots, a_K)$ is positive. By the first condition for each $(a_1, \ldots, a_K)$ there must be at least one such rule, thus $J > 0$. As B_{i_j} is a fuzzy number its support is an interval. Let

$$\text{supp}(B_{i_j}) = [b_{i_j}^-, b_{i_j}^+]$$

In this case the minimum of the membership functions will have the support

$$b^- = \max_{i_j} b_{i_j}^-$$

$$b^+ = \min_{i_j} b_{i_j}^+$$

There is an index i^- such that:

$$b^- = b^-_{i^-}$$

and an i^+ such that:

$$b^+ = b^+_{i^+}$$

Then by applying the second condition to rules i^- and i^+ we have:

$$b^- < b^+$$

Thus the support of the answer is not empty. The opposite clearly holds as well.

Proposition A.7 *If $A = [a^-_1, a^+_1] \times \ldots \times [a^-_K, a^+_K]$ R is a non-degenerate complete numerical rule system on A and $R(a_1, \ldots, a_K)$ is obtained using the mean defuzzification of the response set $B(a_1, \ldots, a_K)$, then it a continuous function on A.*

Proof:
As the rule system is non-degenerate the membership functions $\mu_{A_{i,k}}(a_k)$ are all continuous. The inference is also defined as a continuous function of its arguments thus the DOF for each rule i $nu_i = D_i(a_1, \ldots, a_K)$ is a continuous function of the arguments $a_1, \ldots, a_K$.

For the weighted sum combination method the response function can be written as:

$$R(a_1, \ldots, a_K) = \frac{\sum_{i=1}^{I} \nu_i \beta_i M(B_i)}{\sum_{i=1}^{I} \beta_i \nu_i} \tag{A.0.4}$$

The completeness of the system ensures that

$$\sum_{i=1}^{I} \beta_i \nu_i > 0$$

So the response function is obtained by dividing a continuous function with a nonzero continuous function, thus R is continuous. In the case of the normed weighted sum combination the response function is:

$$R(a_1, \ldots, a_K) = \frac{\sum_{i=1}^{I} \nu_i M(B_i)}{\sum_{i=1}^{I} \nu_i} \tag{A.0.5}$$

The same arguments as for the weighted sum combination prove the continuity for this case.

Let $\nu_i = D_i(a_1, \ldots, a_K)$. For any combination method the response set B can also be written as $B = B(\nu_1, \ldots, \nu_J)$. Let $B' = B(\nu_1, \ldots \nu'_j, \ldots, \nu_J)$. Then, by definition for any combination method and any x, one has

$$|\mu_B(x) - \mu_{B'}(x)| \leq |\nu_j - \nu'_j|$$

Thus for the fuzzy mean $M(B)$ one has:

$$\left| \int_{-\infty}^{+\infty} t\mu_B(t)\, dt - \int_{-\infty}^{+\infty} t\mu_{B'}(t)\, dt \right| \leq$$

$$\leq |\nu_j - \nu'_j| \left(\max_i(\mathrm{supp}(B_i))\min_i(\mathrm{supp}(B_i)) \right) \max_i(\mathrm{supp}(B_i))$$

$$\left| \int_{-\infty}^{+\infty} \mu_B(t)\, dt - \int_{-\infty}^{+\infty} \mu_{B'}(t)\, dt \right| \leq |\nu_j - \nu'_j| \left(\max_i(\mathrm{supp}(B_i))\min_i(\mathrm{supp}(B_i)) \right)$$

Proposition A.8 *If $f(a_1, \ldots, a_K)$ is a continuous function on $\mathcal{A} = [a_1^-, a_1^+] \times \ldots \times [a_K^-, a_K^+]$ then for any $\varepsilon > 0$, any inference combination and any defuzzification method there exists a rule system $\mathcal{R}$ such that*

$$|f(a_1, \ldots, a_K) - R(a_1, \ldots, a_K)| < \varepsilon$$

for each $(a_1, \ldots, a_K)$.

Proof:
The proof follows from the uniform continuity of the function f on $\mathcal{A}$. The proof is only sketched here.

As $f(a_1, \ldots, a_K)$ is continuous on the closed bounded set $\mathcal{A}$ than by the uniform continuity of f for each $\varepsilon > 0$ there is a δ such that

$$|f(a_1, \ldots, a_K) - f(b_1, \ldots, b_K)| < \varepsilon$$

if

$$\|(a_1, \ldots, a_K) - (b_1, \ldots, b_K)\| < \delta$$

Let the rule system be constructed with the help of triangular fuzzy arguments with supports $[a_k^- + \frac{m_1\delta}{2L}, a_k^- + \frac{(m_1+2)\delta}{2L}]$ and the corresponding response being $(f^* - \epsilon, f^*, f^* + \epsilon)_T$ with

$$f^* = f(\ldots, a_k^- + \frac{(m_1 + 1)\delta}{2L}, \ldots)$$

An appropriate choice of L and ϵ leads to the desired result.

Proposition A.9 *Suppose that all rules of a non-degenerate numerical rule system are formulated with arguments $A_{i,k}$ which have differentiable membership functions $\mu_{A_{i,k}}(x)$ for all real x values then using the product inference rules and the mean defuzzification method the resulting function $R(a_1, \ldots, a_K)$ is differentiable.*

Proof:
The membership functions $\mu_{A_{i,k}}(a_k)$ are all differentiable. The inference is also defined as a differentiable function of its arguments thus the DOF for any rule i $\nu_i = D_i(a_1, \ldots, a_K)$ is also a differentiable function of its arguments $a_1, \ldots, a_K$.

In particular for the weighted sum combination method the response function can be written as:

$$R(a_1, \ldots, a_K) = \frac{\sum_{i=1}^{I} \nu_i \beta_i M(B_i)}{\sum_{i=1}^{I} \beta_i \nu_i} \tag{A.0.6}$$

The completeness of the system ensures that

$$\sum_{i=1}^{I} \beta_i \nu_i > 0$$

Since the response function is obtained by dividing a differentiable function with a positive differentiable function, the quotient R is also differentiable. The same argumentation can clearly be used for the normed weighted sum combination method. For the other combination methods the proof is somewhat more difficult.

Index